HISTORICAL GIS

Technologies, Methodologies and Scholarship

Historical GIS is an emerging field that uses Geographical Information Systems (GIS) to research the geographies of the past. Ian Gregory and Paul Ell present the first study to comprehensively define this emerging field, exploring all aspects of using GIS in historical research.

A GIS is a form of database in which every item of data is linked to a spatial location. This technology offers unparalleled opportunities to add insight and rejuvenate historical research through the ability to identify and use the geographical characteristics of data. *Historical GIS* introduces the basic concepts and tools underpinning GIS technology, describing and critically assessing the visualisation, analytical and e-Science methodologies that it enables, and examining key scholarship where GIS has been used to enhance research debates. The result is a clear agenda charting how GIS will develop as one of the most important approaches to scholarship in historical geography.

Ian N. Gregory is a Senior Lecturer in Digital Humanities at Lancaster University.

Paul S. Ell is Founding Director of the Centre for Data Digitisation and Analysis in the School of Geography, Archaeology and Palaeoecology at the Queen's University of Belfast.

CAMBRIDGE STUDIES IN HISTORICAL GEOGRAPHY 39

Series editors

Cambridge Studies in Historical Geography encourages exploration of the philosophies, methodologies and techniques of historical geography, and publishes the results of new research within all branches of the subject. It endeavours to secure the marriage of traditional scholarship with innovative approaches to problems and to sources, aiming in this way to provide a focus for the discipline and to contribute towards its development. The series is an international forum for publication in historical geography, which also promotes contact with workers in cognate disciplines.

Books in the series

1 Period and Place: Research Methods in Historical Geography. *Edited by* Alan R. H. Baker and M. Billinge
2 The Historical Geography of Scotland since 1707: Geographical Aspects of Modernisation. David Turnock
3 Historical Understanding in Geography: an Idealist Approach. Leonard Guelke
4 English Industrial Cities of the Nineteenth Century: a Social Geography. R. J. Dennis*
5 Explorations in Historical Geography: Interpretative Essays. *Edited by* Alan R. H. Baker and Derek Gregory
6 The Tithe Surveys of England and Wales. R. J. P. Kain and H. C. Prince
7 Human Territoriality: its Theory and History. Robert David Sack
8 The West Indies: Patterns of Development, Culture and Environmental Change since 1492. David Watts*
9 The Iconography of Landscape: Essays in the Symbolic Representation, Design and use of Past Environments. *Edited by* Denis Cosgrove and Stephen Daniels*
10 Urban Historical Geography: Recent Progress in Britain and Germany. *Edited by* Dietrich Denecke and Gareth Shaw
11 An Historical Geography of Modern Australia: the Restive Fringe. J. M. Powell*
12 The Sugar-cane Industry: an Historical Geography from its Origins to 1914. J. H. Galloway
13 Poverty, Ethnicity and the American City, 1840–1925: Changing Conceptions of the Slum and Ghetto. David Ward*
14 Peasants, Politicians and Producers: the Organisation of Agriculture in France since 1918. M. C. Cleary
15 The Underdraining of Farmland in England during the Nineteenth Century. A. D. M. Phillips
16 Migration in Colonial Spanish America. *Edited by* David Robinson
17 Urbanising Britain: Essays on Class and Community in the Nineteenth Century. *Edited by* Gerry Kearns and Charles W. J. Withers
18 Ideology and Landscape in Historical Perspective: Essays on the Meanings of some Places in the Past. *Edited by* Alan R. H. Raker and Gideon Biger
19 Power and Pauperism: the Workhouse System, 1834–1884. Felix Driver
20 Trade and Urban Development in Poland: an Economic Geography of Cracow from its Origins to 1795. F. W. Carter
21 An Historical Geography of France. Xavier De Planhol
22 Peasantry to Capitalism: Western Östergötland in the Nineteenth Century. Goran Hoppe and John Langton

23 Agricultural Revolution in England: the Transformation of the Agrarian Economy, 1500–1850. Mark Overton*
24 Marc Bloch, Sociology and Geography: Encountering Changing Disciplines. Susan W. Friedman
25 Land and Society in Edwardian Britain. Brian Short
26 Deciphering Global Epidemics: Analytical Approaches to the Disease Records of World Cities, 1888–1912. Andrew Cliff, Peter Haggett and Matthew Smallman-Raynor*
27 Society in Time and Space: a Geographical Perspective on Change. Robert A. Dodgshon*
28 Fraternity Among the French Peasantry: Sociability and Voluntary Associations in the Loire valley, 1815–1914. Alan R. H. Baker
29 Imperial Visions: Nationalist Imagination and Geographical Expansion in the Russian Far East, 1840–1865. Mark Bassin
30 Hong Kong as a Global Metropolis. David M. Meyer
31 English Seigniorial Agriculture, 1250–1450. Bruce M. S. Campbell
32 Governmentality and the Mastery of Territory in Nineteenth-Century America. Matthew G. Hannah*
33 Geography, Science and National Identity: Scotland since 1520. Charles W. J. Withers
34 Empire Forestry and the Origins of Environmentalism. Gregory Barton
35 Ireland, the Great War and the Geography of Remembrance. Nuala C. Johnson
36 Geography and History: Bridging the Divide. Alan R. H. Baker*
37 Geographies of England: the North–South Divide, Imagined and Material. Alan R. H. Baker and Mark Billinge*
38 White Creole Culture, Politics and Identity during the Age of Abolition. David Lambert
39 Historical GIS: Technologies, Methodologies and Scholarship. Ian N. Gregory and Paul S. Ell*

Titles marked with an asterisk are available in paperback.*

HISTORICAL GIS

Technologies, Methodologies and Scholarship

Ian N. Gregory
Lancaster University

Paul S. Ell
Queen's University of Belfast

CAMBRIDGE
UNIVERSITY PRESS

University Printing House, Cambridge CB2 8BS, United Kingdom

Published in the United States of America by Cambridge University Press, New York

Cambridge University Press is part of the University of Cambridge.

It furthers the University's mission by disseminating knowledge in the pursuit of education, learning and research at the highest international levels of excellence.

www.cambridge.org
Information on this title: www.cambridge.org/9780521671705

First published 2007

A catalogue record for this publication is available from the British Library

Library of Congress Cataloguing in Publication data
Gregory, Ian, 1970–
Historical GIS: technologies, methodologies, and scholarship / Ian N. Gregory and Paul S. Ell.
p. cm. (Cambridge studies in historical geography)
Includes bibliographical references.
ISBN 978-0-521-85563-1 (hardback) ISBN 978-0-521-67170-5 (pbk.)
1. Historical geographic information systems. I. Ell, Paul S. II. Title. III. Series.
G70.212G74 2007
910′.285 – dc22

ISBN 978-0-521-85563-1 Hardback
ISBN 978-0-521-67170-5 Paperback

CONTENTS

FIGURES

TABLES

ACKNOWLEDGEMENTS

This book benefited from funding from a number of research grants, including a Leverhulme Early Career Fellowship awarded to Dr Gregory (ECF40115) and an award to both authors under the ESRC's Research Methods Programme (H333250016). Thanks also to Dr Alastair Pearson (University of Portsmouth) for permission to use the cover illustration. This is taken from his research, and shows tithe rent charges around Newport, in Pembrokeshire, in 1845.

1

GIS and its role in historical research: an introduction

1.1 INTRODUCTION

Until the mid-1990s, most historians or historical geographers would not have heard of a Geographical Information System (GIS), despite its widespread use in other disciplines. Since then there has been a rapid increase in awareness of the potential of GIS in historical research, such that a new field, historical GIS, has emerged. This book will demonstrate that historical GIS has the potential to reinvigorate almost all aspects of historical geography, and indeed bring many historians who would not previously have regarded themselves as geographers into the fold. This is because GIS allows geographical data to be used in ways that are far more powerful than any other approach permits. It provides a toolkit that enables the historian to structure, integrate, manipulate, analyse and display data in ways that are either completely new, or are made significantly easier. Using these tools and techniques allows historians to re-examine radically the way that space is used in the discipline. As with any new approach, enthusiasm for GIS is not without risks. GIS originated in disciplines that use quantitative and scientific approaches in a data-rich environment. Historical geography is rarely data-rich; in fact, data are frequently incomplete and error-prone. As a discipline, historical geography employs both quantitative and qualitative approaches and is rightly suspicious of overly scientific or positivist methodologies. Further, many researchers are introduced to GIS as a software tool rather than an approach to scholarship. Although at its core GIS is software-based, to be used effectively it has to be seen as an approach to representing and handling geographical information that provides scholars with information on both *what* features are and *where* they are located. The researcher using GIS should be asking 'what are the geographical aspects of my research question?' rather than 'what can I do with my dataset using this software?'

This book describes how GIS can be used in historical research. Historical GIS is a highly inter-disciplinary subject combining historical scholarship with expertise in using GIS. GIS stresses the geographical aspects of research questions and datasets. This book focuses on this approach in historical research. As it is written for an audience of historians, historical geographers and others with an interest in the past, it will not say much about traditional historical approaches, which will already be familiar to the reader. The philosophy of the book is that GIS has much to offer to our understanding of history and that it must be used appropriately, critically and innovatively so that the limitations of spatially referenced data are understood, and that GIS is used in ways that are appropriate for historical research.

The first section of this chapter describes the advantages of GIS's ability to handle Geographical Information. The chapter then turns to describing how GIS can be productively used in historical research. The evolution of the use of GIS in geography and the controversies that ensued are then described so that lessons can be learnt by historians. Finally, we look at how GIS is spreading into historical scholarship and how it should be best used.

These are themes that will be further developed as the book progresses. Chapters 2, 3 and 4 introduce the fundamentals of GIS data and tools. Chapter 2 examines how GIS models the world and how this enables and limits what can be done with GIS. Chapter 3 looks at one of the more onerous and arguably under-valued aspects of historical GIS research: creating databases. Chapter 4 then explores the basic tools that GIS offers to the historian. Chapters 5 to 8 build on these basic concepts to show the different approaches to applying GIS to historical research. Chapter 5 evaluates how the mapping and visualisation abilities offered by GIS can allow exploration of the spatial patterns of the past. In Chapter 6 we explore how time can be added to basic GIS functionality to enable the study of temporal change. Chapter 7 goes beyond desktop GIS to examine the potential for data retrieval and integration over the internet. Chapter 8 then looks at how GIS can be used to perform the quantitative analysis of spatial patterns. Finally, in Chapter 9 we provide a critique of the various ways in which GIS is being used by historians following both quantitative and qualitative approaches to advance their understanding of the geographies of the past. In this way we offer an in-depth critique of technologies and methodologies of historical GIS, and explore how these are changing scholarship in historical geography.

1.2 DEFINITIONS: GEOGRAPHICAL INFORMATION AND GEOGRAPHICAL INFORMATION SYSTEMS

Geographical Information Systems (GIS) and its related fields have a maze of definitions that vary between authors and depend heavily on context (Chrisman, 1999).

There are, however, some basic terms related to the subject that are important to understand. The best place to start is by defining Geographical Information (GI). At its broadest, any information that refers to a location on the Earth's surface can be considered as Geographical Information. In practice, this covers almost all information so it is perhaps more helpful to regard information as GI when it includes some information about location which is relevant to the problem at hand. The location of the railways that make up a transport network or the location of the boundaries that define an administrative system are obvious examples of Geographical Information. Census data are also Geographical Information as they consist of statistics on clearly bounded locations on the Earth's surface such as enumeration districts (EDs), census tracts, registration districts or counties. Information on historical monuments or data on hospital mortality rates can also be GI if they include information on where the features are located. Qualitative sources can also be GI. Examples include texts referring to various places, drawings or photographs of buildings around a town, and sound recordings of place names. The lack of a clearly defined location need not mean that information is not GI, although it may limit the uses to which it can be put. For example, information about 'the area around London' or a Chinese province for which no accurate boundaries exist (or may ever have existed) can still be GI.

What then are Geographical Information Systems? The narrowest answer to this is to regard GIS as a type of software. In simple terms, a GIS is a computer package that is designed to represent Geographical Information effectively. It is, therefore, a system that allows us to handle information about the location of features or phenomena on the Earth's surface. This is usually done by combining a database management system (DBMS) with a computer mapping system. Thematic information that says *what* a feature is is stored as a row of data in the DBMS. Technically, this is referred to as *attribute data*. Each row of attribute data is linked to information on *where* the feature is located. This is termed *spatial data* and is stored using co-ordinates but is usually represented graphically using the 'mapping system' (Cowen, 1990). 'Mapping system' is perhaps an over-simplification, as in addition to providing the ability to draw maps, this deals with all the functionality that is explicitly spatial, including responding to queries such as 'what is at this location?' and calculating distances, areas and whether features are connected to each other. A GIS software package is thus a geographical database management system as is shown in Fig. 1.1.

Moving beyond this limited definition involves defining GIS not as a type of software but as the tools that the software offers. A widely quoted definition was produced by a government enquiry into the use of Geographical Information which defined GIS as '[a] system for capturing, storing, checking, integrating, manipulating, analysing

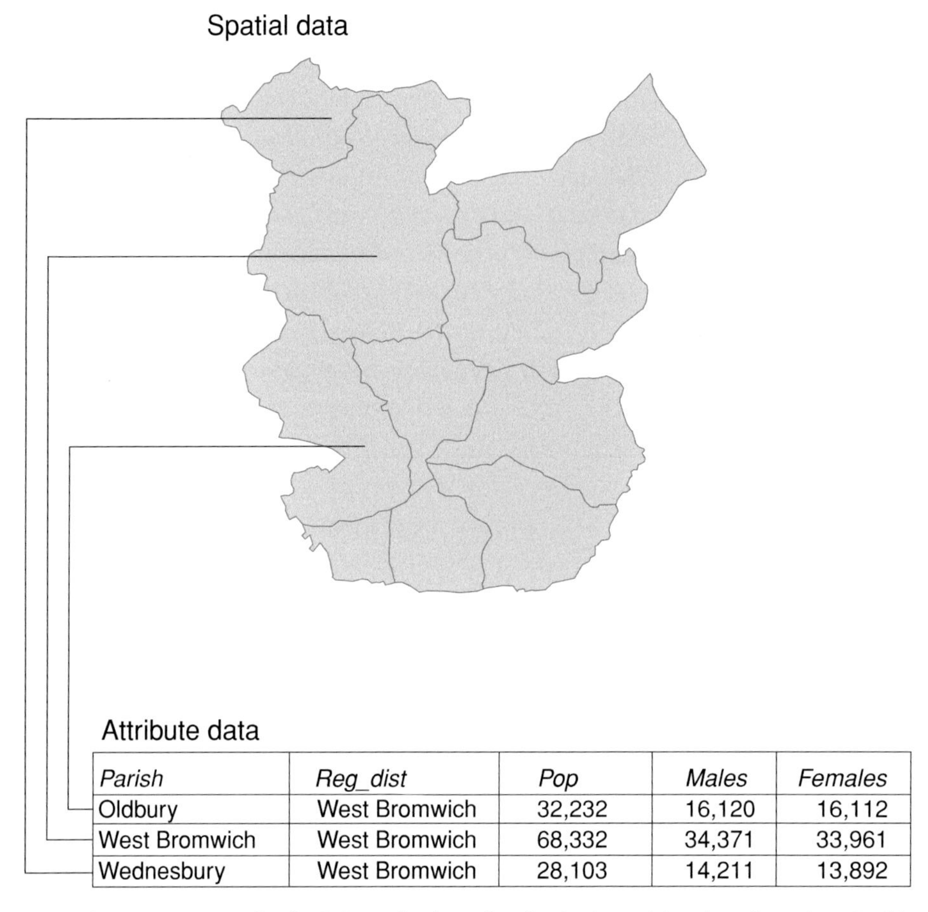

Parish	Reg_dist	Pop	Males	Females
Oldbury	West Bromwich	32,232	16,120	16,112
West Bromwich	West Bromwich	68,332	34,371	33,961
Wednesbury	West Bromwich	28,103	14,211	13,892

Fig. 1.1 An example of a GIS as a database showing both spatial and attribute data. In this example the spatial data represent some English parishes while the attached attribute data are statistical data taken from the census.

and displaying data which are spatially referenced to the Earth' (Department of the Environment, 1987: p. 132). Other authors list slightly different characteristics, but it is broadly agreed that the key abilities of GIS are that it allows a geographical database to be created and the data in it to be manipulated, integrated, analysed and displayed.

Both of these approaches are bottom-up definitions in that they define GIS by looking at the technology or the functionality that it offers. It is also useful to take a top-down approach that asks what using GIS offers to the scholar. Researchers using GIS often talk about 'following a GIS approach'. This involves making use of both components of the data, the spatial and the attribute, to look explicitly at how spatial patterns vary across the Earth's surface. Although it can be argued that this is what geographers have always done, a GIS approach is distinguished by both the power and the limitations of how data are represented in GIS software, something that will be described in detail in Chapter 2. It is important to realise, however, that it is not necessary to use GIS software to follow a GIS approach, although it is usually advantageous to do so.

'Following a GIS approach' effectively must involve conducting high-quality geographical research in ways that are enabled by GIS. It is therefore worth considering briefly what the discipline of geography is concerned with. Johnston (1983) argues that geography is the study of places and the inter-relationships between them. He says that there are two main aspects to the geographer's approach: the vertical, which attempts to study individual places and the components that make them up in detail, and the horizontal, which attempts to study the relationships that link separate locations. It is the ability to do both, he argues, that gives geography its integrity as a discipline. Geography is thus fundamentally concerned with space. Horizontally it is concerned with the flows of people and ideas. Vertically most places can be broken down into increasingly small units to the level of the individuals and households that have distinct locations within a place. Space, therefore, frequently determines the arrangement of how people interact with each other, and with the natural and man-made environment. It is important also to note that a place is rarely, if ever, an un-sub-dividable whole. Instead, it is made up of smaller components that will also have locations. These include, for example, households within a village, or tracts within a city. As GIS is fundamentally concerned with locating data in space, clearly it has much to offer to the researcher who is interested in studying these vertical and horizontal geographical relationships.

As well as applying GIS to academic study, there is a strong research agenda that looks at how best to handle Geographical Information in appropriate and effective ways. This has become known as Geographical Information Science (GISc) (Goodchild, 1992a). The term 'science' is perhaps an unfortunate one from the historian's point of view as it implies a strongly quantitative approach. In fact, there is a significant amount of work in the GISc arena that is relevant to qualitative historical research and where historians have much to offer GI scientists. This is particularly true of how to handle qualitative data in a GIS environment: an area where significant advances have been made by historians, as will be discussed in Chapter 9.

As well as the GIS and GISc arenas, GIS is closely linked to three other research areas: spatial analysis, geocomputation and scientific visualisation. Spatial analysis (variously also referred to as spatial data analysis, spatial statistics, spatial modelling and geographical data analysis[1]) refers to the use of Geographical Information in a statistical environment. It seeks to explore spatial patterns; therefore, unlike with conventional statistical approaches, results will vary depending on where the observations are located (Fotheringham, 1997). One example of spatial analysis is testing the locations of patients suffering from a disease to see if they cluster in certain parts of the study area. Another would be testing to see if high values of one dataset tend to occur near high values of another. Spatial analysis can be done without using GIS – indeed the sub-discipline emerged before GIS – but using the two together has clear advantages as GIS software allows the spatial component of the data to be handled effectively – for example, to calculate distances or to display results through maps. The use of spatial analysis with GIS is returned to in Chapter 8.

Geocomputation is a newer term. Whereas spatial analysis uses the power of statistics to obtain knowledge from geographical data, geocomputation uses computing power and computing science. It has been defined as '[t]he eclectic application of computational methods and techniques "to portray spatial properties, to explain geographical phenomena, and to solve geographical problems"' (Couclelis, 1998: p. 17). Geocomputation emphasises using highly computationally intensive techniques such as artificial intelligence in a high-performance computing environment (Openshaw, 2000). Unlike spatial statistics, geocomputation is a very young sub-discipline whose value in academic research has yet to be fully established.

Scientific visualisation, otherwise known as visualisation in scientific computing (ViSC) (Wood and Brodlie, 1994), involves representing data graphically so that they can be explored and understood. Mapping is an obvious form of scientific visualisation that can be used effectively to explore Geographical Information. Other less explicitly spatial forms of scientific visualisation, such as graphs, are also important ways of visualising Geographical Information. Digital technology allows us to go beyond these traditional forms of visualisation into areas such as interactive maps or diagrams that can be queried and changed, animated maps that show change over time and virtual worlds that allow exploration of a detailed representation of a re-created landscape. All of these techniques are well suited to exploring Geographical Information and complement a GIS approach well (Visvalingham, 1994). Visualisation is returned to in Chapter 5.

GIS software provides a database management system that is capable of representing Geographical Information in ways that make use of both its attribute, which says what the data is, and its location, saying where it is. This 'dual-representational

[1] All of these terms can be defined in slightly different ways, but the differences are too specific to be relevant here.

framework' (Peuquet, 1994) is what distinguishes GIS, and it opens a range of functionality to allow researchers to handle Geographical Information effectively. Using a GIS approach involves making use of both of these components of data simultaneously. As will be discussed later in this chapter, this is not radically new; however, modern computing power means that many of the problems and complexities that have hampered this approach in the past can now be handled. We are thus in a position to incorporate effective use of both components of Geographical Information into historical geography.

1.3 GIS AND THE THREE COMPONENTS OF DATA

The previous section introduced the concept that Geographical Information has two components: the attribute that says what the object is, and the spatial that describes where it is located. In reality, most information has a third component: time.

Geographers and historical geographers as diverse as J. Langton and D. Massey have long argued that to understand a phenomenon fully requires a detailed understanding of all three of its components. Langton (1972) argued this from the perspective of systems theory. He argued for what he termed *diachronic* analysis in which the parameters in the system could be measured as frequently as possible so that the effects of a process could be measured. In this way, inappropriate or unrealistic assumptions such as a system being in equilibrium could be avoided. Implementing this effectively would require large amounts of attribute, spatial and temporal information to measure the impact of change over time and space and thus allow an understanding of the process to be developed. Without detail in all three components, the understanding that could be gained will inevitably be simplistic, heavily reducing the effectiveness of this form of analysis.

More recently, Massey (1999 and 2005) also stressed the importance of an integrated understanding of space and time. She argued that to understand an individual place at a set point in time we need to understand the chronology of how the place developed. This provides a single story or example of how a place develops and thus how a process can operate. To gain a fuller understanding requires multiple stories giving a variety of examples of how different places develop differently. These authors, writing nearly thirty years apart, and from very different perspectives, have a common thread: that gaining understanding requires an understanding of attribute, space and time together.

In an ideal world, information on all three components of the data would be available. Unfortunately, this is not often the case, especially when working with

Table 1.1 *The representation of geographic data in various formats*

	Fixed	Controlled	Measured
Census data	Time	Location	Attribute
Soils data	Time	Attribute	Location
Topographic map	Time	Attribute	Location
Weather reports	Location	Time	Attribute
Flood tables	Location	Time	Attribute
Tide tables	Attribute	Location	Time
Airline schedules	Location	Attribute	Time

Source: adapted from Langran and Chrisman, 1988: p. 2.

historical sources. Langran and Chrisman (1988) argue that in many instances, providing a detailed measurement of one of the three components of data requires the second to be controlled or limited and the third to be completely fixed. One example of this is the census: to measure information about people and households as an attribute accurately, the census arbitrarily subdivides the country into discrete spatial units, to control location, and is taken on a specific night so time is fixed. Soils data provide a different example. Soils are first subdivided into arbitrary classes, controlling attribute, and the survey is again done at a fixed time. The boundaries between the different soil types are recorded, so location is accurately measured. A variety of examples of this type are shown in Table 1.1.

Representing a source provides a second layer of complexity. Again, choices are frequently made that further erode the detail inherent in at least one of the three components. To return to the example of the census, many researchers have analysed the census using a database management system. These have no concept of location beyond place name and administrative hierarchy. As these are extremely crude forms of spatial information, location is no longer simply controlled; it is effectively removed. Less extreme examples include simplifying the spatial component by aggregation or using approximate locations.

The previous section described how GIS enables attribute and location to be handled in an integrated manner. It is not, however, explicitly designed to handle time, although strategies can be developed to work around this, as will be discussed in Chapter 6. This means that GIS has the ability to handle the detail of all three components of the data contained in a source. To illustrate the importance of this, suppose that there is a database of information on historic artefacts such as pieces of pottery. This contains attribute information that describes the pottery, locational information that gives the co-ordinates where the artefact was found and perhaps where it was made, and temporal information on when the pottery was made and perhaps when it was found. Any database could store this information but only GIS is able to handle it effectively. For example, in response to the query 'where has pottery of a certain kind dating from the 1850s been found?', the GIS would map these results and may show, for example, that these were only found in one quite limited area. Repeating the query for later dates may reveal if and how the pottery spread out over time. Even in this simple way, using GIS allows the researcher to make improved use of all three of the components of the data up to the limitations of the source.

Census data provide a more complex example. As described above, a census measures the number of people by controlling location and fixing time. Using GIS allows the attribute and locational components to be handled together, as the GIS will store the attribute information combined with a graphical representation of the administrative units used to publish the data. This allows the exploration of two components of the data rather than just one. Exploring change over time from the census requires multiple censuses to be represented. GIS enables this by integrating data through location and by using maps and spatial analysis as tools for comparison. It does not automatically solve the problems inherent in the sources, such as changes in the questions asked and the administrative areas used, but it does, as will be returned to in later chapters, provide a way into these problems.

A GIS approach, therefore, helps the researcher to make full use of all three components of a source. In theory, it makes no requirements on the approach that the researcher adopts. GIS is applicable to any form of historical scholarship, quantitative or qualitative, as long as the research is concerned with information on theme, location and, if appropriate, time. In practice, however, GIS requires the data to be structured in certain ways that are often not sympathetic to many forms of historical sources, particularly those containing uncertainty, ambiguity and missing data. It can also be argued that GIS is better at handling quantitative, clearly defined sources such as the census, rather than qualitative or vaguer sources such as textual descriptions. This will be returned to in later chapters but, as the book will demonstrate, these 'softer' sources can be used effectively in GIS.

1.4 THE BENEFITS AND LIMITATIONS OF GIS

From the above discussion it should be clear that a GIS is a form of database management system that allows the researcher to handle all three components of information in an integrated manner with a particular emphasis on location. Gregory *et al.* (2001a) group the benefits of using GIS into three categories: the organisation of historical sources, the ability to visualise these sources and the results gained from analysing them, and the ability to perform spatial analysis on the data. None of these are impossible without GIS, but all may be made significantly easier by using it.

Each item of data in a GIS database contains a co-ordinate-based reference to its location on the Earth's surface. This provides a framework for organising a database that has many benefits. Most obviously, it allows the researcher to query the database to ask where objects are and how their location relates to the location of other objects. Co-ordinates are a particularly useful tool for integrating data from diverse sources. To return to the examples above, it may be that a researcher wants to investigate the relationship between pottery and certain population characteristics recorded in the census – for example, to do with occupations. As both the pottery database and the census database have co-ordinate-based spatial data we are able to integrate the two and find which census zone each piece of pottery was found in. It may be that the datasets use different co-ordinate systems or different map projections, but GIS software should provide the tools necessary to standardise these. Once the two have been integrated the researcher can examine the relationship between the two datasets using whatever methodology he or she sees fit. Chapter 4 describes this in detail.

A further use of co-ordinates in data integration is associated with resource discovery. With the growth of the internet there is a wealth of data available online. However, finding appropriate data can be difficult. The use of metadata, data that describes a dataset, has become a standard way of providing information to online catalogues (Green and Bossomaier, 2002). Co-ordinates are a particularly useful form of metadata. For example, a researcher may be interested in a particular part of the country, say the hundred[2] of Eyhorne in Kent. As this is not a common way of referring to this area it is likely that searching for the name 'Eyhorne' will result in many datasets of relevance to this place being missed. If co-ordinates describing the extent of Eyhorne are used instead, the fact that, for example, our pottery database or census data contains co-ordinates lying within this extent can be discovered. This is discussed in Chapter 7.

The second advantage offered by GIS is the ability to visualise data, particularly through mapping. In GIS the map is no longer an end product; it is now a research tool. As soon as the GIS database is created it can be mapped. This means that the spatial patterns within the data can be repeatedly re-explored throughout the research process, greatly enhancing our ability to explore and understand spatial

[2] Hundreds were a type of English administrative unit that was in use until the nineteenth century but have since fallen out of use.

patterns. The GIS can also be used to output maps and other forms of scientific visualisation for publication. One example of the advantages this brings is the ability to produce atlases quickly and effectively. Examples produced in this way include Kennedy *et al.*'s (1999) atlas of the Great Irish Famine, Woods and Shelton's (1997) atlas of mortality in Victorian England and Wales, and Spence's (2000a) atlas of London in the 1690s. Alternatively, visualisations can be produced in electronic form where more flexible types of product can be published at lower cost than on paper. Examples of this include Gatley and Ell's (2000) CD-ROM of census, Poor Law and vital registration statistics; and Ray's (2002a) electronic map of the spatial and temporal patterns of accused and accusers associated with the Salem witchcraft trials. The full scholarly implications of electronic publication are yet to be established but will undoubtedly be significant. Visualisation and GIS is returned to in Chapter 5.

The final advantage offered by GIS is the ability to devise new analytical techniques that make use of the advantages of spatial data while being sympathetic to their limitations. A valid criticism of many statistical techniques used in quantitative geography is that they summarise a pattern using a single statistic or equation. This gives the average relationship, which is not allowed to vary, across the study area. In reality, geographers are likely to be interested in how relationships vary; thus although nominally geographical, these techniques remove the impact of location. To take geography into account, techniques need to emphasise the differences across a study area (Fotheringham, 1997). Taking advantage of the ready availability of co-ordinates in GIS means that techniques can be devised that emphasise spatial variations rather than averaging them away. The ability to map results easily works in tandem with this analytic ability, as variations can be mapped to highlight geographical variations. This is returned to in Chapter 8.

The challenge for historical GIS is to take technologies and methodologies developed in other areas and apply them to historical research questions in a way that provides new insights into the geography of a research question. This must be done in an appropriate way that fuses GIS methodology with high-quality history and, crucially, high-quality geography. The historian wanting to use GIS must not only learn the technical skills of GIS, but must also learn the academic skills of a geographer. This involves thinking about how people interact with each other and their environment, both natural and man-made, and how places interact with each other. This range of skills means that there are many advantages to collaborative research in historical GIS research. This allows individuals to focus on their own particular strengths rather than requiring one individual to be an expert in many fields.

In summary, although GIS software can be thought of as a spatially enabled database, it opens the potential for historians to handle the spatial characteristics of data in new and powerful ways that in turn improve their ability to explore and understand the geographical aspects of their research questions. To add a note of caution, GIS also allows spatial data to be misused, so researchers using GIS must always consider the limitations and problems inherent in spatial data.

1.5 A BRIEF HISTORY OF GIS

In order to understand what GIS is and, in particular, what its limitations are for historical research, it is helpful to have a brief understanding of how GIS developed. Many of GIS's origins came from the military, particularly in the US (Smith, 1992); however, this is not well documented so this section will focus on the civilian development of GIS.

Computerised handling of Geographical Information originated in the early 1960s from two largely independent organisations: the Harvard Laboratory for Computer Graphics and the Canadian GIS (CGIS). The Harvard Laboratory for Computer Graphics was set up in 1965 by Howard Fisher, an architect. Its aim was to develop automated cartographic procedures capable of using line printers to create maps quickly and cheaply. Its major achievement was to produce the mapping package SYMAP that could produce a variety of maps including choropleths[3] and isolines[4]. This was easy to use on the standards of the day and was acquired by over 500 institutions, half of which were universities. Its major success was that it demonstrated the potential for the automated use of cartography and spawned a variety of subsequent programs (Coppock and Rhind, 1991).

At the same time, Roger Tomlinson was creating what is commonly acknowledged as being the first true GIS. The Canadian Government was planning the Canadian Land Inventory (CLI) to map the land capability of settled Canada. Tomlinson persuaded the Government that the best way to achieve this was by using computer technology. By the end of the 1960s this idea had been turned into a working system that held maps and associated attribute information for the whole of Canada (Tomlinson, 1990).

Another development that occurred late in the 1960s was the use of GIS to help in taking and disseminating the census. The US was particularly influential in this with the development of Dual Independent Map Encoding (DIME), a structure that

[3] Maps where each area is shaded according to the value of one of its attributes – for example, a map of unemployment rates for census districts.

[4] Maps based on lines joining points of equal values, for example contour maps or maps of atmospheric pressure.

helped to encode and check areas built up using streets as boundaries, as is usually the case in urban areas in the US (US Bureau of the Census, 1990).

The 1970s saw increases in computing power and in awareness of the importance of environmental issues. Both of these spurred the growth of GIS, with developments spreading to private software companies in North America and Europe. The 1980s saw the launch of ArcInfo, the GIS software package that was to become the industry standard for the next two decades. ArcInfo was developed by Environmental Systems Research Institute (ESRI), a company that was originally a not-for-profit environmental consultancy. They initially developed ArcInfo for in-house use, but were swift to spot its potential, so the company shifted its emphasis to GIS software development. By the late 1980s they were selling over 2,000 ArcInfo licences a year to a global market (Coppock and Rhind, 1991). In universities, the increasing availability of GIS software was leading to an increasing use of GIS in research and education. This was particularly prevalent in physical geography but spread into the more quantitative elements of human geography with the census being a particularly productive research area.

The cost of hardware, the difficulty of using command-line driven computer programs and the cost of data meant that GIS remained a fairly specialised area of academic geography until the mid-1990s. By then, the increasing availability and power of desktop personal computers (PCs) led to the launch of new software that could run on PCs with user-friendly graphical user interfaces. This opened GIS to a far wider audience. The first widely used software package of this sort was MapInfo from the MapInfo corporation. ESRI responded by launching ArcView as a sister product to ArcInfo. ArcView and ArcInfo have since been merged into a single product: ArcGIS. These developments mean that there are no longer major technical and financial barriers to the use of GIS. Consequently, its use has spread into mainstream geography and other disciplines including history. As will be described below, this has been far from uncontroversial, particularly with a backlash against what was seen as GIS's overly quantitative and positivist approach. As the use of GIS has spread, so the approaches used with GIS have diversified, and there has been an increasing realisation that the use of GIS has to build on and add to existing academic paradigms rather than replace them.

1.6 GIS IN GEOGRAPHICAL RESEARCH

When GIS spread into human geography in the 1980s, a pro-GIS lobby was quick to make lavish claims about its potential to revolutionise the subject. The early arguments for GIS typically stressed that it offered a scientific approach to data through the use of computers, and that this would create a single language behind which the discipline of geography could unite (see, for example, Abler, 1987 and 1993; Hall, 1992; Openshaw, 1991a, 1992 and 1997). It was argued that 'GIS would emphasise an holistic view of geography that is broad enough to encompass nearly all geographers and all of geography. At the same time it would offer a means of creating a new scientific look to geography, and confer upon the subject a degree of currency and relevancy that has, arguably, long been missing' (Openshaw, 1991a: p. 626).

At the time this quote was written, GIS software ran on mainframe computers that were only useable by a minority of geographers with a particularly technical background. Most geographers were thus excluded from what was supposed to be an integrating technology. To make matters worse, the people that could use GIS tended to stress its advantages from a quantitative, data-led, scientific perspective that sat uncomfortably with the approaches of many human geographers. The counter-argument was therefore developed that GIS marked a return to the largely discredited approaches of the 1960s, and ignored developments in the discipline from the 1970s and 1980s. Inevitably, an exclusionary technology that made exaggerated claims about the benefits of using an approach with which many geographers disagreed resulted in an equally exaggerated backlash. The arguments against GIS focused on the scientific approach that it was claimed to advocate, and that its development was driven by technology rather than the needs of academic geography. Examples of the types of criticism that were levelled include Taylor (1990: p. 211) claiming that GIS represented 'the very worst sort of positivism', Pickles (1995a) claiming that GIS lacked any epistemological basis, and Curry (1995) arguing that GIS lacked any treatment of ethical, economic or political issues.

There was a certain degree of hysteria from both sides of this argument, although underlying it were serious issues. In particular, if GIS was to become an accepted part of academic geography, what role should it play? A careful examination of the more extreme claims of both the pro- and anti-GIS camps reveals some common concerns. It is noticeable that even Openshaw acknowledged that GIS on its own would not offer complete scientific understanding but would instead only offer insights into patterns and associations found within the data (Openshaw, 1991b). From the other side, Taylor, this time writing with R. J. Johnston, conceded that 'Geographers of all persuasions and all applied agendas are interested in making full use of available information and in doing so as efficiently and as effectively as possible. GIS offers them much . . . They can portray the world in a complexity and detail that their predecessors could hardly have imagined' (Taylor and Johnston, 1995: p. 63).

Since the mid-1990s when this debate came to a head, the impact of GIS has become clearer. GIS is now widely used in academic geography. This may be in part because of a more sensitive use of GIS as a result of the debate described above, but is probably more due to improvements in computer technology. Rather than requiring specialist hardware and impenetrable software, GIS software now runs on desktop PCs and is relatively easy to learn. Academics have been able to adapt the tools offered by GIS for their own purposes, rather than imposing a new form of geography, GIS has allowed researchers to use their geographical information in ways that they deem suitable, based on their own approach to the discipline.

Most British geography departments now include a certain amount of GIS in undergraduate geography degrees; some now run undergraduate degrees largely focused on GIS, and there are several well established GIS masters courses. In America, with a weaker tradition of universities having geography departments, the impact of GIS has arguably been even more significant. On both sides of the Atlantic there are now regular conferences focused on GIS, such as GIS Research UK[5] in the UK and GIScience[6] in the US. There have been major research initiatives in both countries, such as the National Center for Geographical Information and Analysis (NCGIA)[7] and the Center for Spatially Integrated Social Science (CSISS)[8] in the US, and the Regional Research Laboratories (RRLs) in the UK (Chorley and Buxton, 1991; Masser, 1988). As it has become an accepted part of the geographer's toolbox, GIS has moved out of specialised books and conference sessions, first to papers with subtitles such as 'A GIS approach', and then to simply being mentioned as part of the research methodology. This demonstrates that GIS has truly arrived as an accepted part of the discipline, but makes it difficult to make statements about how widespread its use is.

The debate about the role of GIS in geography is continuing but has softened. Criticisms of GIS now tend to be constructive and focus on generating a debate about improving its effectiveness as a tool and an approach (Pickles, 1999). It has been recognised that a valid criticism of GIS is that it has focused on the more quantitative and empiricist ends of geography but that there is significant potential for GIS in the aspects of the discipline more concerned with social theory (Johnston, 1999).

1.7 GIS IN HISTORICAL RESEARCH

The use of GIS in historical research, or historical GIS as it has become known, is still in its early stages but is developing rapidly. It is generally recognised that the first moves in the field occurred in the mid-1990s (see, for example, Knowles, 2002a)

[5] www.geo.ed.ac.uk/gisruk/gisruk.html. Viewed 28 May 2007.
[6] www.giscience.org. Viewed 28 May 2007. [7] www.ncgia.ucsb.edu. Viewed 28 May 2007.
[8] www.csiss.org. Viewed 28 May 2007.

with Goerke's (1994) *Coordinates for Historical Maps* being arguably the first book in the field. Initial growth was slow, such that Ogborn (1999), writing a review of work in historical geography in 1997, made no mention of historical GIS. Since then there has been rapid change, and Holdsworth (2002) referred to historical GIS as being one of the emerging trends in historical geography. A number of publications have appeared focusing specifically on historical GIS. In 2000, a special edition of a journal, *Social Science History* (vol. 24: 3), was the first to be devoted to historical GIS. This was followed by the publication of an edited book devoted to historical GIS (Knowles, 2002b) and the History Data Service's *Guide to Good Practice in GIS* (Gregory, 2003) as well as a special edition of the journal *History and Computing* (vol. 13: 1). 2005 saw another special edition of a journal, *Historical Geography* (vol. 33) dedicated to historical GIS. These publications, together with a significant but dispersed literature in other journals, indicate the rapid growth in historical GIS. However, their content reveals that many projects using GIS are still in their early stages and that, as yet, there is no commonly accepted understanding of what GIS should offer to history and what its limitations are.

Several authors have attempted to describe the advantages of GIS in historical research. MacDonald and Black (2000) stress its usefulness in providing context for, in their case, the study of the spread of books, by allowing a wide variety of types of data including numerical, textual and visual sources to be brought together in their appropriate location in time and space. Knowles (2002a) argues that GIS allows familiar evidence to be re-examined using the new tools that GIS provides, so that long-standing interpretations can be challenged. Other authors take a more top-down approach. They argue that human beings and their activities are inherently spatial; therefore, to understand human history we need a spatial approach, something that has often been overlooked (Siebert, 2000). The papers in Woollard (2000) provide examples of some of the advantages of this approach.

There has also been some speculation as to what GIS will do to the discipline of historical geography. Baker (2003) argues that the use of maps allows GIS to challenge existing theories or orthodoxies. In fact, the potential for GIS goes far beyond mapping to cover a range of advantages offered by the researcher being able to explicitly locate his or her data in space. Healey and Stamp (2000) argue that GIS can bridge the gap between what they term the traditional historical geography approach that stresses detailed geographical context and other approaches that use detailed temporal or thematic information. Knowles (2002a) contends that GIS is already reinvigorating historical geography in the US which, she argues, was in danger of dying out. She also emphasises the interdisciplinary nature of GIS in bringing together historians, geographers and GI scientists because '[u]sing GIS intelligently requires a grounding in geographical knowledge. Applying the technology to history requires knowing how to conceptualise and interpret historical sources' (2002a: p. xii).

To date, there has not been much discussion of the limitations or drawbacks of GIS in historical research. This is probably as a result of the lack of exposure of historical GIS to the broader community. Knowles (2000) and Siebert (2000) both point out that the cost of setting up a GIS project is one of its major disadvantages. Beyond this practical disadvantage there are also a variety of other limitations that must be considered. Guelke (1997) argues that historical geographers tend to only understand history as 'time' or 'the past' but that history should be seen as the study of human consciousness and the relationships between people. If Guelke is right in this then it seems likely that GIS will in fact strengthen the chronological view of history that he opposes.

Along with others, Pickles (1999) observes that GIS is better suited to some types of data than others. Consequently, work in historical GIS is likely to emphasise sources and subjects that are well suited to the way that it represents the world. This may lead to other sources and subjects being neglected. There are three specific problems that limit GIS's application to historical research. In all cases, research is being done to attempt to resolve these. Firstly, historical sources are often incomplete, inaccurate or ambiguous. The basic GIS data model insists on giving the appearance of a high degree of accuracy in its co-ordinate data although, as will be discussed in Chapter 4, there is a considerable amount of research in the GI science community on ways of handling what are termed 'error' and 'uncertainty'. It should also be noted that these are not necessarily problems with GIS; they are in fact limitations with the sources. The ability of GIS to manipulate the sources in new ways highlights these limitations. It is also important to consider that although the GI science community is exploring solutions to these issues, historians have a long tradition of dealing with them and should not be afraid to put their own solutions forward. The second problem is that GIS does not currently explicitly handle time. As will be described in Chapter 6, strategies can be devised to deal with this. The third problem is that GIS is still better at handling quantitative data than qualitative. This obviously reflects its origins in the Earth sciences, and historians have much to offer to the GI science community with the developments that they are making in this field.

What should historical GIS as a sub-discipline of historical geography be? There is no point trying to be prescriptive in answering this question while the subject is developing rapidly and new ideas are constantly evolving. There are, however, some lessons that can be learned from geography and some basic principles that people using GIS should adhere to. Firstly, it is to be hoped that there should be no exaggerated claims about either the benefits or drawbacks of GIS such as those that occurred in geography in the early 1990s. There should be no talk of a 'GIS revolution' (Openshaw and Clark, 1996) in historical geography; instead GIS needs to evolve into the discipline in ways that build GIS into the approaches currently used in historical geography. GIS approaches should be adapted to the requirements of historical data and historical scholarship and should not be used uncritically.

A more positive approach as to what GIS has to offer is simply that it should encourage historians to think more carefully about geography and the impact and importance of location. This goes beyond thinking visually (Krygier, 1997; Tufte, 1990) to thinking about what impact location and space have on all aspects of human behaviour. As Healey and Stamp (2000: p. 584) say: 'Not only does geography matter, but GIS makes it much easier to determine the precise extent to which it matters in varying locations.' At the same time, we should not over-focus on the impact of geography at the expense of either thematic or temporal information. It is also important to remember the limitations and complexities of using spatial data such as spatial dependence and ecological fallacy described in Chapter 8.

A final point related to this is that historians using GIS should use all of the detail available in their sources to take advantage of GIS's ability to handle all three components of data. In particular, the practice of arbitrarily aggregating data spatially simply to reduce its complexity should be avoided. Wherever possible, data should be used at the level of aggregation it was published at and the impact of this aggregation should be considered in the results. If any of the components of the data must be simplified then the impact of this simplification must also be considered in interpreting the results. Imagination should also be used to bring together sources that may previously have been considered incompatible. Caution does, however, need to be used to ensure that the ways that they are integrated are methodologically sound.

1.8 CONCLUSIONS

The fundamental ability of GIS compared to any other form of database is that it allows location to be explicitly included as part of the data. This significantly improves our ability to understand the geographical aspects of a research question in three ways: firstly, the ability to structure data using location enhances our ability to explore spatial relationships within or between datasets; secondly, it makes mapping and other forms of visualisation far more accessible than they have been traditionally; and thirdly, it provides the ability to perform more formal analyses of the data where the results take into account the importance of location. In short, GIS provides a set of tools that should reinvigorate geographical enquiry by allowing researchers to better handle the complexity of geographical data. The capability to handle location also allows us to improve our understanding of temporal change by comparing spatial patterns over time.

After a controversial start, GIS has become widely accepted in many areas of geography. Historical geography has been relatively slow to adopt the technology and the approach. A major reason for this is the perception that GIS is a quantitative tool. When it was heavily associated with a quantitative, scientific approach, there were justifications for this suspicion. However, as GIS has developed it is becoming increasingly clear that it can be used with imprecise, qualitative sources. This makes it a far more applicable approach in historical geography as it allows it to be used in both the quantitative and qualitative elements of the discipline.

This is not to say that the use of GIS is problem-free. There are significant costs associated with buying GIS software and learning how to use it. Acquiring data, either by purchasing it or by capturing it yourself, can also be expensive. Even once these building blocks are in place there are still many complexities to be faced. Making appropriate use of Geographical Information to gain understanding of a research question requires intellectual skills as well as the technical skills required to use the software. Maps frequently suggest patterns or relationships but are rarely capable of identifying the processes causing them. On a more conceptual level, using GIS may impose a chronological view of history based on certain types of data that are well suited to its data model. This may be at odds with the approach that many historical geographers like to adopt. It is also important to note that GIS is not applicable to all forms of research in historical geography, as it requires the data to be modelled in a particular way, described in detail in Chapter 2.

Although in its early stages, there is a growing understanding of the potential for using GIS. It has the capacity to allow researchers to open up new areas of geographical enquiry and to re-open areas where the complexity of data has traditionally hampered progress. A difficulty in the development of GIS in historical geography is that it requires expertise in conducting historical research, expertise in conducting geographical research, and expertise in using spatially referenced data within a GIS environment. At present, few people possess all of these skills. These disciplinary boundaries present significant obstacles to progress, obstacles that this book aims to resolve.

2

GIS: a framework for representing the Earth's surface

2.1 INTRODUCTION

This chapter describes the basic methods by which a Geographical Information System represents the world. It is necessarily technical and descriptive; however, a good understanding of the underlying principles is essential to provide an understanding of the strengths and limitations of GIS and its utility in historical research. In this chapter we introduce key terms associated with GIS before proceeding to discuss the range of data models that underlie all GIS data and their utility to the historian.

As was identified in Chapter 1, a Geographical Information System can be thought of as a form of database. What makes it unique is that the GIS combines attribute data, which describe the object, with spatial data, which say where the object is located. Traditionally, attribute data would have been statistical or textual; however, more recently, images, sound, video and other multi-media formats are used. Attribute and spatial data are stored together in a GIS *layer* (or coverage or theme). A layer is the rough equivalent of a database table. Each layer holds information about a particular theme or topic combining both the spatial and attribute data.

Within a layer, spatial data are always represented using co-ordinates. These co-ordinates may represent a *point*, a *line*, or an area or zone, known in GIS terminology as a *polygon*. Points, lines and polygons represent discrete features on the Earth's surface, and a GIS based on using these features uses what is termed a *vector data model*. An example of features in a vector GIS might include a database of churches where each church's location is represented by a point and its attribute data is an image of the church such as a digital photograph. A second example could be a database of railway lines, represented by lines, with attribute data such as the line owner and the volume of traffic per day. Finally, census data provide an example of polygon data where the spatial data are districts or tracts represented by polygons, with the accompanying statistical data as attribute data. Some features such as towns or cities may be represented by either a point or a polygon depending on the scale and purpose of the database.

To represent a continuous feature such as land use or relief (height above sea-level) a *raster data model* may be used. In this the Earth's surface is usually subdivided into a regular grid of small *pixels* with each pixel having values attached to it to represent, for example, land-use type or height. As the mesh of pixels is small, the raster data model provides a way of representing a continuous surface rather than the discrete features found in a vector model.

Regardless of whether raster or vector data are used, the two-pronged data model that combines spatial and attribute data allows us to look at what features are represented in the database and where they are located. It is this that opens up the advantages of GIS described in Chapter 1, namely the ability to use location to structure a database and explore its geography, to integrate data from different sources, to visualise data through maps and virtual worlds and to perform spatial analysis on the data (Gregory *et al.*, 2001a). Features such as churches, railway lines or census zones are examples of types of features that can be effectively handled using GIS. As long as accurate locations can be found for them, GIS provides an integrated way of handling information about what a feature is, together with information on where it is, that no other approach, be it on paper or in a database, can rival.

The data model also fundamentally limits what GIS can achieve in two main ways. Firstly, where features have accurate co-ordinate locations that can be effectively represented using points, lines, polygons or pixels, GIS represents them well. Many geographical features are more complex than this. Even a simple example such as a place name may not be well represented by a GIS. For example, if we have some information about 'Bristol', does this refer to the urban limits of Bristol, an administrative district of the same name, or is it best represented by a point whose location approximates to the centre of the city? This becomes even more complicated with cultural regions such as 'the south-west of England', 'the industrial north-east', or 'Welsh-speaking areas'. As it is impossible to assign an accurate location to these, GIS is limited in how it can handle the data. Fundamentally it must resort to using a point, a polygon or a form of raster surface to represent these complex areas. Methods can be developed to handle this spatial uncertainty and these will be described in later chapters. The second limitation is that while GIS handles attribute and location well, the third component of data – time – is not explicitly incorporated into the GIS data model. As described in Chapter 6, methods can be developed to handle this; nevertheless GIS does fall somewhat short of providing an integrated representation of all three components of data (Peuquet, 1994).

The remainder of this chapter will describe in more detail how GIS represents the world, concentrating on spatial data, the differences between the raster and vector data models, how spatial and attribute data are brought together, the use of triangulated irregular networks as an alternative approach to representing a continuous surface, and how layers can be used to bring data together from multiple sources.

2.2 REPRESENTING LOCATIONS USING SPATIAL DATA

As attribute data are likely to be familiar to most readers, they will not be dealt with in any detail here. Further information on attribute databases in GIS can be found in, for example, Heywood *et al.* (2002) or Worboys (1999). Harvey and Press (1996) provide a detailed examination of the use of databases in historical research which includes examples of the types of attribute databases that could be used within a GIS. This section will only deal with spatial data, section 2.4 will return to using spatial and attribute data together.

As introduced above, there are two basic types of spatial data: vector data which use points, lines and polygons to represent discrete features on the Earth's surface, and raster data in which regular tessellations, usually square pixels, represent continually varying features on the Earth's surface. All good introductory guides to GIS describe these data models in some detail – see, for example, Chrisman (2002), Clarke (1997), Jones (1997) or Martin (1996a) – so only a brief description will be provided here.

2.2.1 *Vector data*

In vector data the most fundamental unit of storage is the point. A point is stored in the GIS software as a single pair of (x,y) co-ordinates. This represents a location on the Earth's surface which, in theory, has neither length nor width. A point is usually presented to the user graphically and its co-ordinates are kept hidden. Figure 2.1 gives a simplified representation of how points are stored as spatial data in a GIS.

A line is created by joining a string of individual points. These points, often referred to as *vertices*, are placed where the line changes direction, a straight line being drawn by the software between each point. A line must consist of at least two points: the start and end points, usually referred to as *nodes*. Wherever a line has a junction, the GIS will store this as a node. Lines between nodes are often referred to as *segments* or *arcs*. Figure 2.2 shows a simplified example of how lines are stored in a GIS. It should be noted that the choice of where vertices are placed is somewhat subjective and will

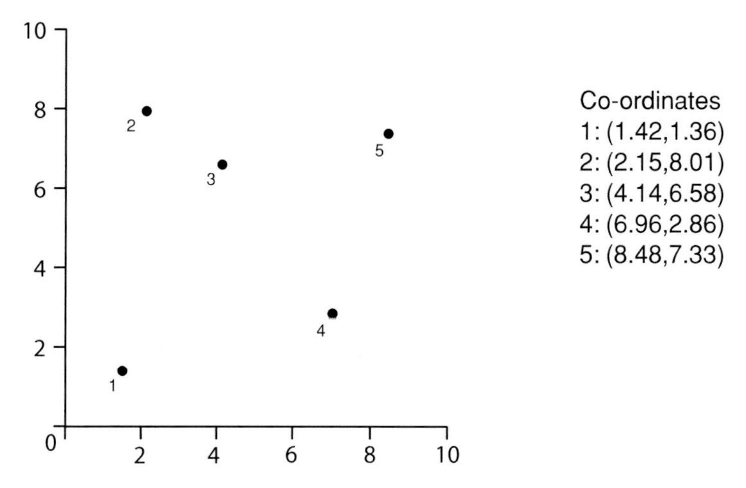

Fig. 2.1 Spatial data: points. Points are represented by a single pair of (x,y) co-ordinates.

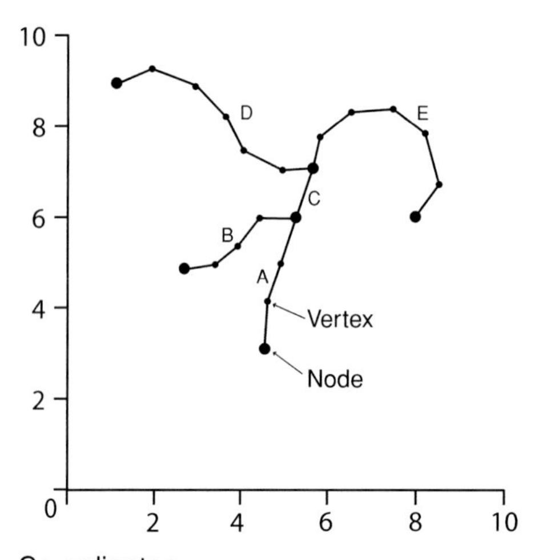

Co-ordinates
A: (4.58,3.11) (4.56,4.09) (4.92,4.97) (5.95,5.98)
B: (2.67,4.85) (3.41,4.98) (3.89,5.33) (4.45,5.97) (5.95,5.98)
C: (5.95,5.98) (5.65,7.11)
D: (5.65,7.11) ... (1.12,8.94)
E: (5.65,7.11) ... (8.03,5.98)

Fig. 2.2 Spatial data: simplified representation of lines in a GIS. Lines consist of a series of straight lines between points. The start and end points of a line are termed *nodes*, the intermediate points are termed *vertices*. The diagram shows five line segments labelled *A* to *E*. *A* starts at node (4.58,3.11) and finishes at node (5.95,5.98) with two vertices in-between. *B* starts at node (2.67,4.85) and, like *A*, finishes at node (5.95,5.98) with three intervening vertices. *C* starts at node (5.95,5.98), the same node that *A* and *B* finish with, and finishes at node (5.65,7.11). As it is a straight line it has no intervening vertices. For simplicity, the co-ordinates of the vertices on lines *D* and *E* are not shown.

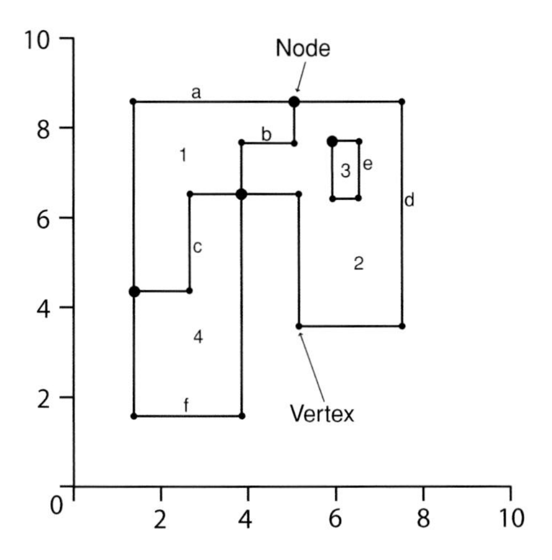

Polygons – line segments
1: a,b,c
2: b,d,e
3: e
4: c,f

Lines – co-ordinates
a: (1.36,4.38) (1.36,8.62) (5.12,8.62)
b: (5.12,8.62) (5.12,7.65) (4.03,7.65) (4.03,6.52)
...
f: (1.36,4.38) (1.36,1.55) (4.03,1.55) (4.03,6.52)

Fig. 2.3 Spatial data: simplified representation of polygons in a GIS. Polygons are created by completely enclosing an area using one or more line segments. Polygon *1* is completely enclosed by line segments *a*, *b*, and *c*; polygon *2* by *b*, *d*, and *e*; and polygon *3* just by *e*. Line segments are represented in the same way as in Fig. 2.2, thus line segment *a* consists of two nodes and a single vertex, line *b* of two nodes and two vertices, and so on.

depend on the scale of the source map the data were captured from (see Chapter 3). A straight line only requires two points: one at the start and one at the end. Line *C* in fig. 2.2 is an example of a straight line segment. A complex line such as a rocky coastline will require many points. Smooth curves such as those found on railway lines or contours are more subjective, but usually require a large number of points so that the curve does not have sudden changes of direction. The length of a line segment can be measured, but, in theory, lines have no width.

Finally, polygons are created by completely enclosing an area using one or more line segments. A polygon therefore is two-dimensional, having both length and width. Most GIS software packages can calculate both the area of a polygon and the length of its perimeter. A simplified method of storing polygons is shown in fig. 2.3. Each line segment, sometimes referred to as a *polyline*, is stored in the same way as the line segments described above. For each polygon the GIS stores the ID number of the line segments that enclose it, thus polygon *1* in fig. 2.3 is enclosed by line segments *a*, *b* and *c*. Polygon *3*, on the other hand, is completely enclosed by line *e*. An advantage of this structure is that a line is frequently a boundary between two polygons, such as line *b* being the boundary between polygons *1* and *2*. In this structure the computer only needs a single representation of the line. This both reduces file space required and has the advantage that when the data are captured (entered into the computer; see Chapter 3), each line need only be captured once.

This introduces a highly important concept with major implications for the functionality of the GIS, namely *topology*. Topology is concerned with the connectivity, adjacency and location of features in relation to each other. It provides additional information to co-ordinates which give absolute locations, shapes, distances and directions. The London Underground map is a good example of a topological map that does not give information on absolute locations. It makes it easy to calculate which lines to travel on to go from one station to another but there is no information on the shapes of the lines (for example where they go round bends) and the only information that it gives on the distances and directions are by implication rather than accurately measurable. The topological structure used in a polygon layer means that for each polygon the GIS knows the line segments that form its boundaries, and for every line segment it knows which polygons lie on either side. This means that polygon data have a clear structure in which the GIS knows which line segments make up each polygon's boundaries and, from here, which polygons are neighbours. It also means that it is logically impossible for a location to belong to more than one polygon (Egenhofer and Herring, 1991; Peuquet, 1990). This has obvious advantages when the study area is clearly subdivided, for example into administrative districts, but means that if there is genuine ambiguity in the real world there is no way that the GIS can easily represent this.

Lines can also have a topological structure. Figure 2.4 shows an example of how this is implemented. The location of each junction between lines is stored as a node. For every node the identification of each line segment that connects to it is stored. This means that the system knows which line segment connects to each other line segment. Thus, rather than having an unstructured collection of lines, we now have a *network*. This allows the GIS to calculate routes from one location on the network to another. If we wanted to go from node *1* on fig. 2.4 to node *7*, the system is able to use this structure to calculate how this can be done. One route would involve following segment *A* to node *2*, taking segment *E* to node *4*, and then following segment *G* to reach the destination at node *7*. An alternative route would involve going from

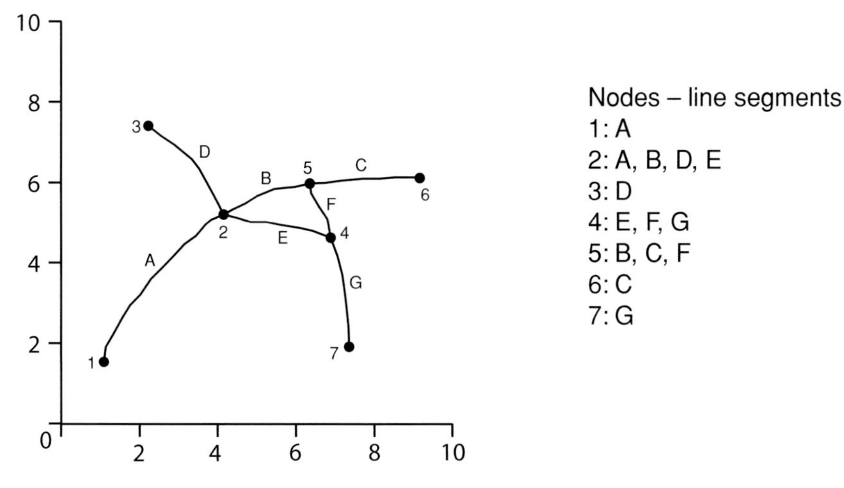

Fig. 2.4 Simplified representation of forming a network from line segments and topological information. For each node an ID number and the line segments that connect to it are stored. This allows the computer to calculate which line segments it needs to travel along to get from one node to another. The co-ordinates of line segments would be stored as shown in Fig. 2.2.

segment *A* onto segment *B* at node *2*, then going onto segment *F* at node *5*, and going onto segment *G* at node *4* to reach the destination. Computationally this is complex and it has been kept deliberately simple for this example; however, these are the basic principals that underlie vehicle routing software packages that allow a user to calculate the shortest or quickest route between two points on a road or railway network.

In vector GISs there may, therefore, be two forms of locational information. The co-ordinate-based information that is present in all spatial data gives the absolute location of each feature on the Earth's surface. The topological information that may be present gives information on the location of features in relation to each other and on how different features connect to each other. It is possible to have vector lines or polygons without topological information being present, in which case it is known as *spaghetti data*, reflecting the fact that it is unstructured. Usually, however, having both co-ordinate and topological data provides information on both where features are and how they connect to each other.

2.2.2 *Raster data*

Raster data are conceptually quite different. The study area is subdivided into a regular grid with each square (or other shape) representing a small part of the

1	1	1	1	4	4	1	1	1	1	1	1	1	1	1
1	1	1	1	1	4	1	1	1	1	1	1	1	1	1
5	1	1	1	1	4	4	2	2	2	2	1	1	1	1
5	5	5	1	1	2	4	4	2	2	2	2	2	1	1
5	5	5	5	2	2	4	4	2	2	2	2	2	1	1
5	5	5	5	2	2	4	4	2	2	1	1	1	4	4
5	5	5	1	1	1	4	4	2	1	1	4	4	1	1
5	5	1	1	1	1	1	4	4	4	4	1	1	1	1
5	1	1	1	1	1	1	4	4	1	1	1	1	1	1
4	4	4	4	4	4	4	1	4	4	1	1	1	1	1
1	1	1	1	1	1	1	1	4	4	2	1	1	1	1
3	3	1	1	1	1	1	3	2	4	4	2	2	1	1
3	3	3	3	3	1	3	3	1	2	4	4	2	1	1
3	3	3	3	3	3	3	3	1	1	4	4	1	1	1
3	3	3	3	3	3	3	3	3	1	1	4	4	1	1

Legend
1: Agricultural
2: Urban
3: Forest
4: Road
5: Water

Fig. 2.5 A raster surface in a GIS. A simplified representation of land use as represented by a raster surface. Each pixel is allocated to one land-use class.

Earth's surface. In the GIS each tessellation is represented by an individual pixel. In theory a pixel can be almost any shape; however, in reality squares are almost always used (DeMers, 2002; Egenhofer and Herring, 1991). Unlike in a vector system, the boundaries between pixels are not intended to represent any meaningful boundary on the ground. Instead, pixels are so small that they can be assumed to be internally homogenous and allow the feature being represented to vary continuously over space.

Figure 2.5 shows an example of a raster GIS representing land use. In reality the pixels will be far smaller than those used on the diagram. Each pixel has a numeric value that, in this case, represents its land use. The raster system has no concept of lines or polygons; therefore, although there is a thick road running from north to south and a thinner road running from west to east, the system has no built-in concept that these are lines. In places, due to the arrangement of the pixels, the line may not appear continuous, or the thick line may appear to be thinner. There is also no concept that there are continuous polygons of urban land or water. It is possible to find out which neighbours a pixel has, but using this to define lines can be difficult as there may well be unexpected breaks in lines caused by the way the grid breaks up the underlying surface. Also, in a raster system a co-ordinate is not explicitly attached to each feature. Instead, the co-ordinates of one part of the layer, frequently the bottom-left-hand corner, are recorded, as well as the dimensions of each pixel. In this way the location of each pixel can be calculated but is not explicitly stored.

2.2.3 *Comparing raster and vector data*

There are, therefore, both conceptual and technical differences between vector and raster systems. The basic difference is that a vector system stores information about discrete objects and gives them precise locations which, in the case of polygons, will have unambiguous boundaries. The raster system stores a continuous representation of the Earth's surface which does not have such precise locations, does not require unambiguous boundaries, but does not have concepts such as lines.

The question of which to use depends on the data to be included in the GIS and the purpose for which the system is to be used. One general rule that is frequently quoted is that vector systems are more suitable for human features, while raster systems are better for physical or environmental systems. This is over-simplistic but does have a certain basis. Another suggestion is that vector systems are more suited to precise and well-known features while raster systems are better where there is inherent uncertainty in the data. A historical example of this means that where researchers have created GISs of nineteenth and twentieth-century census data, such as the Great Britain Historical GIS (GBHGIS) (Gregory *et al.*, 2002) or the Quantitative Databank of Belgian Territorial Structure (De Moor and Wiedemann, 2001), they have used vector systems as these are well suited to representing the precisely defined and well-mapped administrative boundaries used to publish these sources. As a contrast, where Bartley and Campbell (1997) created a GIS of land use and land value in pre-Black Death England using medieval records taken from wills, the *Inquisitions Post Mortem* (IPMs), they used a raster system in part because there was no way from their sources that they could place clear boundaries between estates or land types. These projects are all discussed in more detail later in the book.

Another way of thinking about the difference between raster and vector is to consider the example of a historical map that we want to use to create a GIS. The map is a conventional topographical map whose features include roads, rivers, settlements, forests and administrative boundaries as well as place names, and hill shading to help represent relief. There is also cartographic information such as a legend, a scale and a date. Methods of capturing the data from the map will be discussed in the next chapter; however, to illustrate the difference between raster and vector, this provides a useful example. A raster data model would provide a single representation of the map that would be identical to the original, apart from two changes: the electronic version will be degraded by error introduced by the digitising process and will be enhanced because a real-world co-ordinate system will be added to the raster scan. Effectively, in the raster version, each pixel will have a value that represents its colour on the original map. This allows hill shading to be encoded and place-name information to remain interpretable to a human user. It has the advantage of providing context but may cause problems, as a place name that lies across a road will break the road line and may mean that automated procedures are not able to identify that the road is a continuous line. The raster model would have pixels representing the administrative boundaries but no concept that they enclose areas such as parishes or districts. This would be further complicated if dotted or dashed lines are used on the source map as it becomes almost impossible for a computer system to identify these as continuous lines.

A vector representation of the map would be more abstract than the raster. Each type of information, such as roads, rivers, land use and administrative areas, would be stored in a separate layer of data (see section 2.5 for more information on layers); for example, the roads would be stored on a line layer, and the administrative areas as a polygon layer. A vector system can then identify routes along roads, that there are distinct administrative units, which neighbours they have, and so on. Lengths of lines and areas of administrative units are easily calculated. Much of the contextual information is, however, removed. The hill shading is not readily representable in the vector system. Methods could be developed to handle the place names; however, they would not be a facsimile of their representation on the source map. This may cause problems, as the original cartographer may have placed names carefully along a line or used well-spaced large letters to represent large areas of the map. If we zoom in on a line in the raster system, the line will get larger, as zooming in is done by increasing the pixel size. In the vector system the line will remain the same width, as the vector data model has no concept of the width of a line. Cartographers often use line width to handle uncertainty, as it disguises exactly where a thick line is placed. The raster system maintains this but in the vector system it is lost. Thus, the vector system is more abstract and has more functionality as it has topology, but will lose some of the contextual information. The raster system is a more direct representation of the original source. This is returned to in section 2.5.

2.3 CREATING DIGITAL TERRAIN MODELS

An alternative method of representing the Earth's surface is through the use of a triangular irregular network (TIN). Although technically these are a form of vector data, they are quite different to other models. TINs are usually used to represent relief. They often form the basis of digital terrain models (DTMs), sometimes called digital elevation models (DEMs), which attempt to create a three-dimensional representation of the landscape. The process of creating a DTM using a TIN starts with point data for which in additional to (x,y) values, we also have a z value representing height. This is shown in fig. 2.6a where we have the locations of a series of spot heights. If a line is drawn from the point with height 11 near the north-west of the diagram, to the point at height 60 slightly to the south-east, the length of this line can be calculated using the co-ordinates of the two points. As the difference in the heights of the two points is known, it is possible to calculate the slope of the line. If we did the same from the point at height 11 to point 5 to the south-west, this slope can also be calculated and will be shallower than the previous one.

A TIN uses this idea. Every point is joined to every other point using a straight line; where lines intersect the longer one is removed. This gives the effect shown in fig. 2.6b where the spot-heights have been converted into a TIN. Not only does this provide the length and angle of every line, it also subdivides the study area into triangles and for each triangle the gradient and aspect (angle that it faces) can be calculated. Thus, the triangle created from points 11, 60 and 5 will have an approximately westerly aspect; that by 60, 2 and 10 a south-easterly aspect, and so on. Once we have a TIN this can be converted into a three-dimensional (or more correctly 2.5-dimensional) representation of the landscape. Figure 2.6c represents this by using the TIN to create contours. The areas between the contours have been shaded to give an impression of the relief of the area.

A DTM is thus created using a TIN which in turn is created from a collection of spot heights, based on the assumption that there is a regular change in gradient between each spot height (Burrough and McDonnell, 1998; Weibel and Heller, 1991). If we have a dense collection of spot heights, this can create a very realistic representation of the Earth's surface, incorporating additional information on slopes and aspects. Although usually used to represent relief, in theory any data for which there are x, y and z values could be used to create a DTM. DTMs have been created of rainfall from rain gauges at specific locations where the amount of rainfall is used as the z value, and even of human patterns such as population where, for example, towns are represented as points with their total population becoming the z value. The validity of doing this does, of course, depend on the validity of the assumption that the distribution of the phenomena of interest changes smoothly between the spot heights. This is likely to be a realistic assumption for rainfall but is less likely to be valid for population data. DTMs can also be created from a regular grid of spot heights such as may be provided as raster data, or from contour lines.

Once created, DTMs offer a way of representing the Earth's surface that has previously been unavailable without creating a three-dimensional model of the landscape. Often, once a DTM has been created, a layer of raster data is draped over it to provide a representation of the features on the Earth's surface. Rumsey and Williams

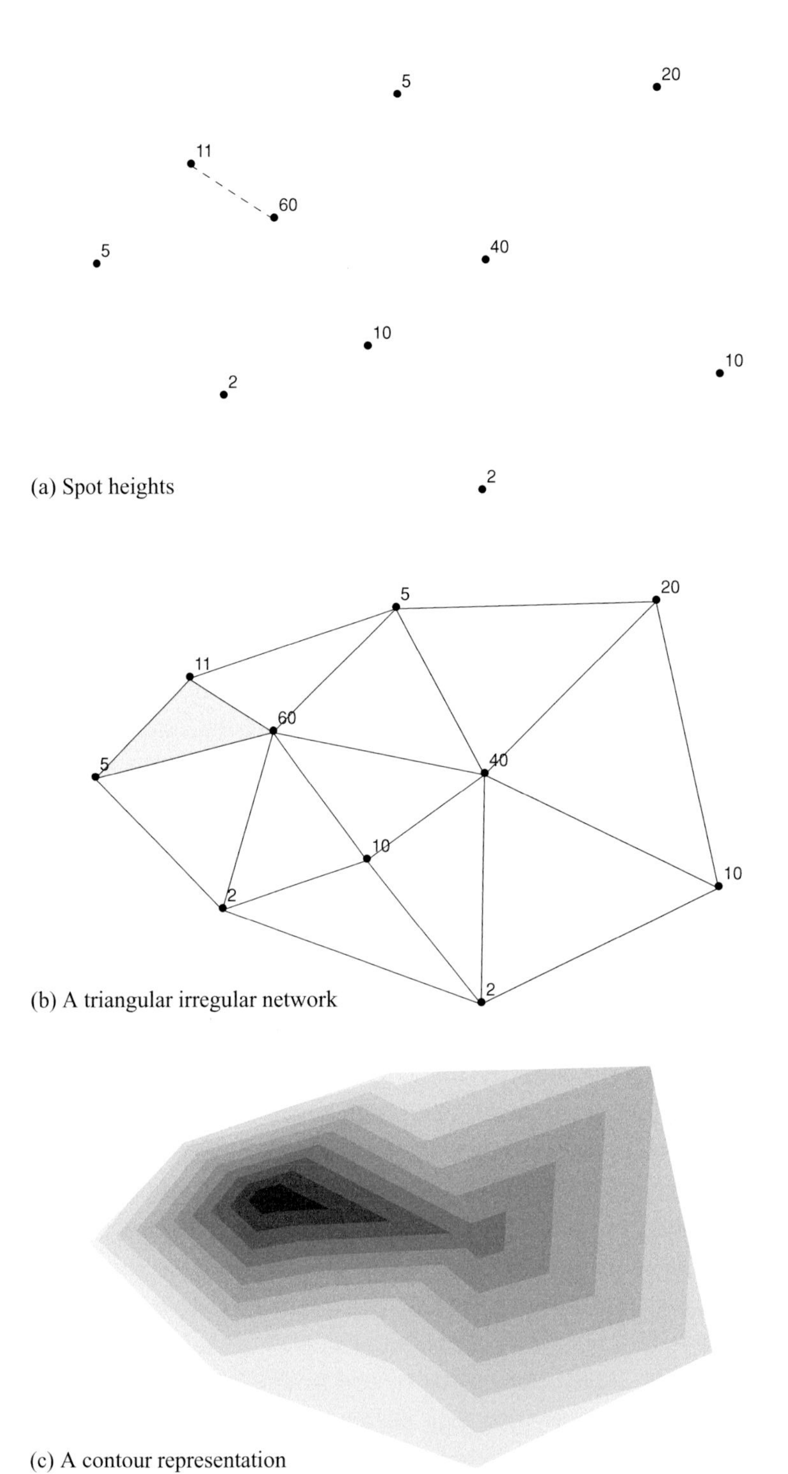

Fig. 2.6 Digital terrain modelling using a triangulated irregular network. (a) A collection of points representing spot heights with height as a *z* value. (b) The spot heights are joined by straight lines to create a TIN. (c) The resulting TIN is used to create a model of the terrain. In this diagram contours have been interpolated (estimated) and higher areas shaded using increasingly dark colours.

(2002) provide an example of this. Their main interest is in historical maps. To bring these to life they create a DTM of San Francisco and drape a 1915 map over it to give an impression of the importance of topography to the city's development. They also take a modern bathymetric TIN (one showing only underwater variations in height) and drape a 1926 map that includes depth markings over this. This demonstrates vividly the difficulties in spanning the Golden Gate and also shows clearly how landfill between 1926 and the present has affected the underwater landscape.

Harris (2002) provides a more artistic drape over a terrain model to try to re-create how the area around an ancient burial mound in Ohio, located in an area that is now heavily urban and industrial, might have looked when the mound was built around 2,500 BC. The DTM is again created from modern data but a drape that includes correct vegetation for the time identified using pollen records is included, as is the impact of sunlight which can be calculated from the aspects in the underlying TIN. This is described in more detail in Chapter 5.

Pearson and Collier (1998) use a DTM to generate extra variables for a statistical analysis. They are interested in agricultural productivity in mid-nineteenth-century Wales and have a variety of data from the census and tithe surveys on population, land ownership and land use. By creating a DTM they are able to add variables on altitude, slope and aspect to their model. This is described in more detail later in this chapter.

A DTM can also be used to perform new forms of analysis that would previously have been impossible. A type of analysis called *viewshed analysis* (Weibel and Heller, 1991) can be used that enables the user to identify all the areas that would have been visible from a particular point on a DTM. Knowles (2004) uses this approach in her analysis of the Battle of Gettysburg to try to get a clearer understanding of the decision-making process of commanders on that day. Using a DTM and viewshed analysis she is able to provide new insights into what a commander could and could not see at crucial moments during the battle.

2.4 SPATIAL DATA AND ATTRIBUTE DATA COMBINED

So far, this chapter has concentrated on spatial data; however, it is the ability of GIS to handle spatial and attribute data together that gives it much of its power. As introduced in section 2.1, attribute data, sometimes referred to as *non-spatial data*, are simply data as most people understand them. They can be any kind of data from statistics to multi-media. Most people coming to GIS will already have an attribute database that they want to use within the GIS and this will usually be

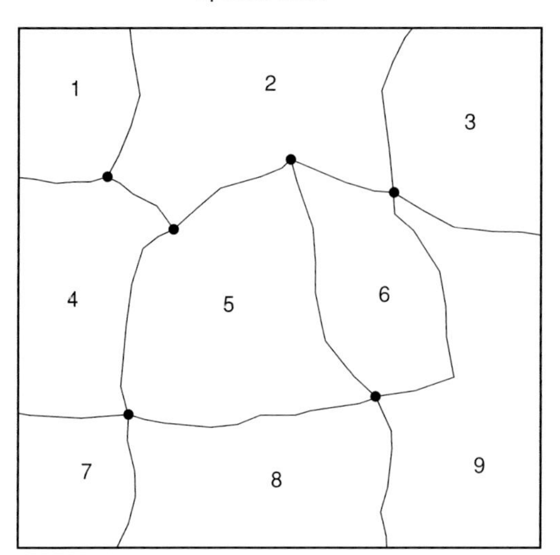

Attribute data

ID	Name	Population
1	Framfield	928
2	Heathfield	746
3	Bramber	1,284
4	Wartling	1,021
5	Twineham	1,675
6	Newick	1,739
7	Ardingley	690
8	Dallington	1,351
9	Isfield	871

Fig. 2.7 Spatial and attribute data. In this example the attribute data (on the right) are linked to polygon spatial data (on the left) using ID numbers.

stored in a database management system such as Oracle, Microsoft Access or Dbase. One approach to dealing with this would be to export all of the attribute data from the DBMS and import it into the attribute database of the GIS software. This is usually not necessary as most GIS software systems allow direct linking from their internal attribute-database management systems to external systems using ODBC (open database connectivity) connections or sometimes through proprietary software. Generally, it is a good idea to store data in the package that handles it best. GIS software is written to handle GIS operations; other database management systems will handle other operations more effectively, while yet more packages such as statistical software may be required for other operations. It is thus a mistake to think that because a researcher is doing GIS-based research they should keep all of their data within the GIS software. Pearson and Collier (1998; see also Pearson and Collier, 2002) refer to the difference between GIS *sensu stricto*, which is the individual components of the GIS software system, and GIS *sensu lato*, where a variety of software packages are used together. These can include database packages, spreadsheets, statistical packages and so on, in addition to the GIS software. Each package is used for the job it does best.

In this section we will mainly examine how the researcher uses spatial data and attribute data together within the GIS software. Figure 2.7 shows how a polygon GIS combines spatial and attribute data. The attribute data are shown on the right hand side. It is in a conventional database format where there are rows and columns of data. Each row refers to a specific feature in the database while each column contains a specific variable about that feature. The example shown might be some census data where the columns include a place name and the total population of each district. Within the GIS, each row of attribute data is given an ID number that links it to the spatial data. This is usually done internally by the software without the user needing to be aware of it. Thus, on the left-hand side of fig. 2.7 we have the spatial data. In this case these are polygons, but points and lines are handled in the same way.

This gives what Peuquet (1994) calls a 'dual representational framework'. It allows us to ask two types of queries: 'where does this occur?' and those that ask 'what is at this location?' More specifically, a 'where does this occur?' query could be implemented by selecting a row of data within the attribute database, such as row 2, Heathfield, in fig. 2.7. The GIS will show us that this refers to the area covered by polygon *2* in the spatial database. If we perform a query such as 'select all rows of data with a population greater than 1,500' then polygons that satisfy this criterion will be highlighted in the spatial data; thus we see that the two central polygons (numbers *5* and *6*) will be selected. This is known as *attribute querying*, as we are selecting features according to their attribute values. Queries that ask 'what is at this location?' are referred to as *spatial queries*. These are typically done graphically. A user may click on polygon number *3* in fig. 2.7 and the attribute database will return that this is 'Bramber' and has a population of 1,284. Spatial and attribute querying is returned to in Chapter 4.

Conventional relational database technology provides a concept called a *relational join*. In this a row of data from one table appears to be joined to a row of data in a second table through a common column of data that contains identical values. The column or columns that link the two tables are called the *key*. Relational joins are also used in GIS and can be very useful, both for joining data tables within the GIS and for linking data about places held within the GIS software to attribute data held in an external DBMS (Healey, 1991; Worboys, 1999). Figure 2.8 gives an example of this. The GIS software contains a layer of point data representing towns which have place names as attributes. The external DBMS contains two tables: one called *pop_data* that contains information on the total population of each town, and one called *mort_data* that contains information on mortality in the town. All three tables have place names in them, and with this acting as a key, a relational join can link them together to show that 'Oving' has a total population of 1,107, has 13 deaths and is at point *1*; or that point *4*, 'Patcham', is the only place with more than 10.0 deaths per 1,000 people. In this way disparate data can be brought together to provide new information. Most software packages provide this functionality in a user-friendly manner.

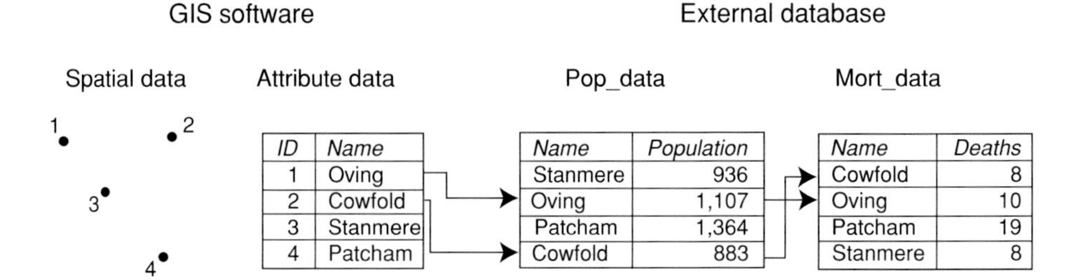

Fig. 2.8 Using relational joins to link to data in an attribute database. A point layer representing villages has names as attributes. Using relational joins with *name* as the key allows us to link this to two tables in an external database; *pop_data*, that holds information about the total population, and *mort_data*, that holds information on mortality. From this we can determine that point *1* had 10 deaths out of a population of 1,107, or that point *4* was the only place to have a death rate of over 10.0 per 1,000 people.

2.5 BRINGING IT TOGETHER WITH LAYERS

As stated above, a layer in GIS is analogous to a table in a database, it is the basic unit of storage for information on one particular subject. It consists of both spatial and attribute data combined in the manner outlined in the previous section. Usually, in GIS each layer is stored in a separate file. In ArcView and ArcGIS the most common file format is the *shapefile* which has become a *de facto* standard for GIS storage. In ArcInfo the *coverage* is used. Confusingly these 'files' typically consist of more than one file. A shapefile, for example, consists of a file with a *.shp* extension to its name, for example mydata.shp, which is the shapefile that the user opens but that only stores the spatial data. A further file has a *.dbf* extension, for example mydata.dbf, and this holds the attribute data in DBase format. If the user opens the *.shp* file in ArcView or ArcGIS and edits the attribute data, the changes will be saved back to the *.dbf* file. A user can also edit the *.dbf* file directly, although this is potentially dangerous as it may corrupt the internal information used by the software to link between the two files. In addition, there may also be files with extensions such as *.sbn*, *.sbx* and *.shx*. These are internal files that contain, for example, indexes and they should not be changed by the user. Thus, a single layer is stored in a single shapefile, but the shapefile actually consists of a number of different files. If any of these files are lost, because they have been deleted or were not copied from one computer to another, the data will be corrupted. The ArcInfo coverage format is even more complex as it stores topology in a relational structure while shapefiles only calculate it when it is required. A readable description of the coverage data structure that explains why this complexity is necessary is provided by Anon (1990).

A layer usually only stores one type of spatial data, be it points, lines, polygons, a raster grid or a TIN. It will also usually only store information about one subject, although this is sometimes a subjective choice; for example, if we want to create a point layer of public buildings such as churches, museums and pubs, it may be desirable to create a single layer containing all public buildings with type as an attribute, or it may be desirable to create a separate layer for each type. If we want to encode the information from a topographic map in a GIS as described earlier, we might have a raster layer that provides a scan of the map itself, a point layer or layers containing the public buildings, line layers containing the roads, railways and rivers, and polygon layers containing the settlement outlines, the administrative units, lakes and woodland. Attributes may be added to all of these containing, for example, the types of public buildings, the classification of the roads, the names of the rivers, the names and type of the administrative units, and the names and types of woodland. A simplified example of how this might be implemented is shown in fig. 2.9.

The real power of layers does not come from their ability to represent a single map; it comes from the ability to integrate data from different sources. Each layer is underlain by a co-ordinate system. Provided this is a real-world co-ordinate such as the British National Grid, Universal Transverse Mercator (UTM), or latitude and longitude, the data from different layers can be brought together. More detail on co-ordinate systems and changing between them is given in Chapter 3, and more detail on the practicalities and issues associated with data integration are given in Chapter 4. For now it is simply enough to consider that if we know the locations of one set of features, we can then determine how these features relate to a second set for which we also have locations.

For a historian this is potentially invaluable. Typically, historical research involves taking information from a number of different sources and trying to bring these together to gain a better understanding of the phenomena under study. As much information has a spatial reference, GIS offers the potential to bring together information from different sources in ways that would previously have been impossible or at least impractical. To demonstrate this, let us return to Pearson and Collier's (1998 and 2002) study of agricultural productivity in Wales in the mid-nineteenth century in more detail. The core of the study was only concerned with a single parish: Newport in Pembrokeshire. To study agricultural productivity they clearly need information on both environmental and human factors. Their key source on human factors was the tithe survey of 1845. This was a detailed record of land use and value in England and Wales that included information on every field, together with maps of the field boundaries. Pearson and Collier created a polygon layer of the fields in Newport parish using a combination of the original tithe mapping and

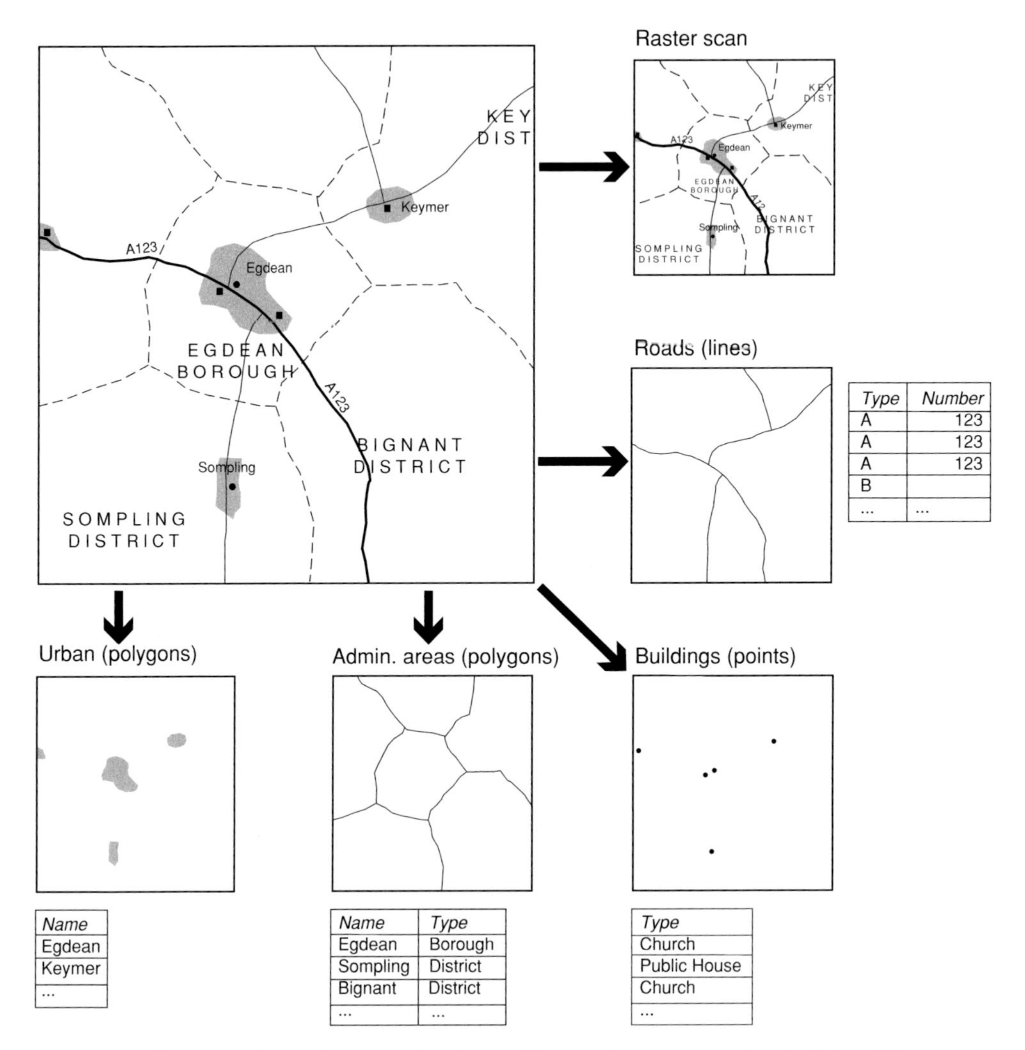

Fig. 2.9 Abstracting a map into layers. To store the paper map (top left) in a GIS, a number of layers are created. These include a raster scan that creates a facsimile of the map, a line layer representing roads with type and number as attributes, a point layer representing buildings with type as attribute, and so on.

a modern 1:10,000 map that showed field boundaries as they exist today. As these boundaries have barely changed over the past 150 years, using modern data provided a more accurate and convenient source of field boundaries than the tithe maps. Using the tithe survey they added attribute data including the name of the owner of the field; the name of the occupant of the fields, as much of the farming was done by tenants; the land-use type such as arable, pasture or meadow; and information on the valuation of the field.

They used two key sources on environmental information. The first was a soils map providing information on the types of soils and the second was data on relief produced by the Ordnance Survey as a raster grid. This was converted into a TIN, and a DTM was produced that provided information on altitude, slope and aspect. The soils and relief data were both modern sources but both can be assumed to have changed little over the preceding 150 years. Bringing this information together meant that, for each field, they had information on: the owner, the tenant, the land use, the tithe valuation, the soil type, the altitude, the slope and the aspect. This gives a far more comprehensive model of factors that may have been affecting agricultural productivity in the mid-nineteenth century than would have been available from a single source. Without the ability of GIS to integrate information by location, it would have been impossible, or at the very least impractical, to have combined all of the information from these three sources at the spatial level of the individual field.

Another example of using layers to integrate disparate historical sources is provided by MacDonald and Black's (2000) study of book history, which was concerned with the spread of literacy in Canada in the nineteenth century. This is a broader project that takes information from a wide variety of sources, including information on transport routes taken from historical maps, demographic information from censuses, and information on libraries, book stores and book production taken from sources such as trade directories. Again, as long as they are able to find a location for each feature, they are able to combine this information and build up an integrated model of the world as it existed in the past. This study is returned to in Chapter 4.

2.6 CONCLUSIONS

GIS is thus a form of database where every item of conventional attribute data is linked to a co-ordinate based representation of its location. The spatial data can be one of four types of graphic primitive: a point, a line, a polygon or a pixel. This model facilitates much of what is described in the remainder of the book, including structuring, integrating, visualising and spatially analysing data. It also fundamentally limits what GIS can achieve. Points, lines, polygons and pixels are crude representations of geographic space. They are well suited to certain types of features such as buildings, roads and administrative units. They are far less suited for less precisely defined features such as cultural regions or spheres of influence. They also rely on a co-ordinate system that allows no imprecision or uncertainty. A GIS will calculate co-ordinates usually to fractions of millimetres and present this as the actual location. It has no explicit way of saying that a feature exists 'near here' or 'at approximately this location'. Both of these problems are likely to affect historians who are used to dealing with data that is not well suited to clearly defined locations, either because it is a phenomenon that does not by definition have a fixed, hard location, or because uncertainty or ambiguity exists as to exactly where a location was. Strategies can be developed to handle these issues, and these will be discussed in later chapters, but they are not explicitly included in the core GIS data model.

A second problem with the way that GIS represents the world is that although it covers attribute and space, it does not explicitly include time. Again strategies for dealing with this will be outlined later in the book (see Chapter 6) but these are not always ideal. A fundamental problem with the vector GIS data model is that there is no way of having consistent topology over time as well as space, thus handling changing polygon boundaries or a changing network is difficult.

In addition, there is a more fundamental problem with the way that GIS represents the world. The 'geographical' part of GIS is spatial data that are created using Euclidean geometry, where space is made up of co-ordinates, straight lines and angles, all of which can be precisely measured (Gatrell, 1983 or 1991). Geography is about far more than the study of Euclidean space; it is the study of places and the interactions between them (Johnston 1983; see also Chapter 1). While some of this can be effectively handled through GIS and its Euclidean data models, much of geography cannot. While GIS has an enormous potential to reinvigorate historical geography (Baker, 2003; Ell and Gregory, 2001; Fleming and Lowerre, 2004; Holdsworth, 2002 and 2003; Knowles, 2005a), there is far more to geography than simply the representations of reality provided by GIS. Therefore, people researching historical geography should consider the geographical implications of their research question, in the broadest sense, and not just the spatial implications as easily handled within GIS. In other words, as Chapter 1 discussed, using historical GIS effectively requires a combination of good history, good geography and good GIS. The GIS enables the researcher to handle Geographical Information in new ways; however, it has the tendency to exclude data that cannot be represented as points, lines, polygons or pixels.

3

Building historical GIS databases

3.1 INTRODUCTION

This chapter examines the issues associated with building historical GIS databases. The process of GIS *data capture*, as it is known, is slow, expensive and frequently tedious. It is almost always the most expensive stage of any GIS project. Bernhardsen (1999) estimates that this stage usually accounts for 60 to 80 per cent of the total cost; other authors present similar figures. The data-capture phase of the project is also time-consuming, and during it there is little obvious short-term reward. This means that, before attempting to build a GIS database, the researcher needs to think carefully about whether the rewards of the completed GIS justify the costs of building it, and about the project management implications of creating the GIS, the complexity of which is frequently underestimated by academics. It is also important to remember that once a GIS database has been built, it will often have uses that go far beyond the original purpose that it was created for. For this reason, it is also important that the researcher considers issues such as documentation, metadata and long-term preservation to ensure that the data can be used appropriately by others.

As discussed in Chapter 2, GIS data are structured using either a vector or a raster data model. Vector data use points, lines and polygons to represent discrete features on the Earth's surface, while raster data represents a continuous surface using pixels. The two types of data are captured in very different ways. Vector data are captured using a process known as *digitising*, which in GIS has a very specific meaning. Digitising involves clicking on points or lines using a *puck* (see section 3.2 below) or cursor to capture their co-ordinates. Raster data are captured by scanning, in a similar way that a photograph or document might be scanned. In both cases, the resulting data are then *geo-referenced* to convert their raw co-ordinates into a real-world co-ordinate system, such as the British National Grid or latitude and longitude.

In GIS terminology, capturing data from a map or another source, such as aerial photographs, is termed *secondary data capture* because the source is at least one stage removed from the real world. The alternative to secondary data capture, *primary data capture*, is to take data directly from the real world. There are a variety of sources of primary data. The Global Positioning System (GPS) may be used to survey features directly. This is often done in archaeological applications. Remotely sensed data, such as satellite images, that provide a digital representation of the Earth's surface are also primary sources. The GIS definition of primary and secondary sources is therefore very different to that used by historians. A map produced in the 1870s is a primary source to a historian if it is a historical document produced in the period that he or she is studying, but is a secondary source in GIS terms because the information that it contains has already been abstracted from the Earth's surface. In this book, the GIS terminology will be used; therefore, a secondary source is typically a historical map.

Historians are used to the idea that the limitation of a source will inherently limit what can be done with the source. Map-based data present additional challenges. In particular, the scale of a map fundamentally limits the accuracy of the features shown and the detail to which they are shown. Once the data are in a GIS, it becomes possible to zoom in on the map in ways that are impossible on paper, and it can appear that scale is no longer an issue. This is not the case, as the digital representation is, at best, of equal quality to the source, but will usually be of lower quality due to inaccuracy and error introduced in the digitising process. This means that the limitations of scale must be considered and, in particular, the balance between using the most detailed data available to get the most accurate representation possible and the extra costs that this will involve.

This chapter deals with the practicalities of digitising and scanning data from secondary sources to produce vector and raster data respectively, geo-referencing data, creating primary data, capturing attribute data, and documenting data to ensure that others can use it. While the costs and difficulties of building a historical GIS database are stressed and should not be underestimated, it is also well worth considering the scholarship involved. Typically, building a GIS involves bringing together data from different sources and integrating these together to create a structure that allows new knowledge to be gained. Although this has tedious aspects, it is also a considerable scholarly undertaking that requires the historian to be familiar with the historical and the geographical implications of the sources used, and that produces a resource that should be made available to scholars for years to come and can be used in ways that the original creator perhaps never even envisaged (Healey and Stamp, 2000).

3.2 CAPTURING DATA FROM SECONDARY SOURCES

The majority of data capture for historical research will be done from secondary sources. This will typically be historical maps, although other sources such as areal photographs should not be overlooked. Data capture is likely to degrade a source, but the image may subsequently be enhanced using operations such as image processing described below. The digital version can be used in ways that go far beyond the uses that the original data were created for. It is therefore important to create the best possible representation of the original, and also to document details of the source. Standards of accuracy for the data capture should be included as part of the documentation. This information is commonly referred to as *data capture metadata* and, where many similar images are being captured through image scanning, it can often be applied automatically, as the same data-capture procedure is replicated for each image. This should ensure that the digital version created becomes a resource that can be used by many future historians.

A map is converted into raster data by scanning. A scanner is passed over the source and produces a representation of it by breaking it down into small pixels. Each pixel is given a value or series of values to represent its colour. A number of different types of scanner are available, each with their own advantages and disadvantages. *Flatbed scanners* are the most common. These are not unlike photocopiers, in that the document is placed face down on a pane of glass and is then scanned. This produces very good results if the document can be placed completely flat on the scanning glass. For bound documents, this may not be possible without compromising the binding. In this instance, a *book-page scanner* can be used that cradles the document – which would normally be bound – face upwards, and an overhead scanner captures an image of the page. This has the advantage that it is unlikely to damage the original, as the source document does not have to be pressed flat onto a glass scanning surface. The disadvantage is that the image quality is unlikely to be as good, particularly as the scanned image will reflect the curvature of the bound page. Both flatbed and book-page scanners are unlikely to be sufficiently large to capture documents exceeding A1 size. This is a concern in GIS where the researcher may wish to scan large map sheets. In this case, the use of a *drum scanner* is recommended. This involves taping the document to a large cylinder. In the scanning process, the cylinder rotates the document over the scanner, rather than the scanner moving over the document.

When scanning, the choice of resolution, or pixel size, is critical. Pixel size is usually expressed in dots per square inch (dpi), which gives the number of pixels captured from every square inch of the page. The higher the resolution of the scan, or the larger the number of dpi, the more detail is captured from the original source. The disadvantage of using a high resolution is that it requires a more expensive scanner, takes longer and produces larger files. It should be remembered, however, that in the long-term, large file sizes are becoming less of a problem as disk space, memory, processing power and network bandwidth become cheaper. One solution that may be appropriate to deal with the problem of large files is to scan at a higher resolution than is required or practical at present, and to keep this version as an archival file. A lower-resolution copy of this file is then made – most image processing software allows this to be done easily – and this version is then used for dissemination over the internet, for example. This is particularly sensible when scanning objects which are rare or are likely to be damaged by the scanning process. Under these circumstances, it is best practice to use the best quality image possible, perhaps up to 3,200 dpi, so that the scan is fit for all likely future uses, and rescanning will not be necessary.

Just as different pixel resolutions can be used, different levels of colour can also be used. Some scanners may only offer *bi-tonal* – black and white – image capture, although *grey-scale* scanning is more common. When colour scanning, the colour bit-level determines the range of colours that it is possible to capture in the scanned image. A higher colour bit-level gives a more accurate representation of the colours being captured, but also increases the file size. Using 8-bit colours, a maximum of 256 different shades are available; using 16-bit increases this to 65,536, and 24-bit and 36-bit increase this to many millions. To achieve the best copy, and thus avoid the risk of having to re-scan in future, 24-bit or 36-bit true-colour scans should usually be made.

After the initial choices about equipment, resolutions and bit-levels have been made, scanning is a fairly automated process. Digitising is altogether more labour intensive. Traditionally, digitising required the use of either a *digitising table* or *digitising tablet*. These are basically the same, the difference being that a tablet is usually smaller, perhaps up to A3 size, while a table is larger, being designed to accommodate entire map sheets. The technology of the two is the same. The table or tablet's surface has a fine mesh of wires underlying it that generate a magnetic field. The map is stuck down to the surface so that it is completely flat. The operator then uses a *puck* to capture the data. The puck is a hand-held device that has a number of buttons and a fine cross-hair. The cross-hair is placed over a point on the map and a button is pressed. In this way, the co-ordinates of the location where the button was pressed are captured and stored to the computer. The co-ordinates are usually in digitiser inches: inches from the bottom left-hand corner of the table or tablet. Lines are stored by moving the puck along the length of the line, pressing the button whenever the line shows a significant change in direction. Thus a vertex is captured every time the button is pressed. Different buttons on the puck are usually used to capture nodes at the start and end of the lines. In this way, a digitiser is well suited to capturing vector data either as points or as lines. As discussed in Chapter 2, polygons are created by completely enclosing an area using one or more lines.

More recently, it has become possible to digitise vector data without using a digitising table or tablet, with a technique called *head-up digitising*. This involves scanning the map, having a software package draw it on a computer screen and then using the cursor to click on the points on-screen, in the same way that a puck is used on a table or tablet. This has the advantage of meaning that an integrated database is created whereby the final product has both a representation of the source map as a raster scan, plus layers of vector data that have been abstracted from it as was illustrated in Fig. 2.9. It is also possible to enhance the scanned image using image-processing software, making, for example, roads and administrative boundaries clearer, allowing them to be more easily plotted using the cursor. Whether using a table, a tablet or head-up digitising, digitising is a slow process, requiring a large amount of manual effort and care by the operator.

Significant effort has been made to develop techniques to automatically extract vector data from a raster scan. This is known as *raster-to-vector conversion* and would lead to considerable savings in terms of the time taken to perform head-up digitising. Unfortunately, these efforts have not had much success, as, although identifying lines from a map may appear to be a simple task, it is hard to develop software that will do it in an automated manner; therefore, the operator frequently has to make subjective choices and value judgements. For example, where a letter from a place name lies over a road, an operator will know to continue the line over the letter, even digitising points in approximately the right locations, if required. For software, to make these kinds of judgements is very difficult. Even more complicated is where lines are shown on a map using dots or dashes. To a human, these do not present any particular challenges, but developing software capable of interpreting this has proved difficult. Therefore, although there have been products developed that do allow a certain amount of raster-to-vector conversion, they tend to be expensive and require a large amount of operator intervention to do an accurate job.

Vector-to-raster conversion is computationally easier, although usually less useful for GIS. It is often used by software to convert data from one or more vector layers to an output map in a bitmap graphics format, but may also be used to create a raster GIS layer from a vector layer.

Scanning and digitising are both prone to error from a number of sources. The term *error* simply means the difference between the real world and the digital representation of it (see Chapter 4). This is seldom simply mistakes, but derives instead from inaccuracies, simplifications and incompleteness caused by the limitations of the source and the way in which the data were captured. The original source's limitations are primarily concerned with the scale, the features included and excluded, and with accuracy of the original survey and subsequent damage to the source. As discussed above, scale determines how much detail can be used to show features. Accuracy refers to how effective the original survey and printing were in placing real-world features onto their correct location on the paper map. Any inaccuracy in smaller-scale maps will be amplified by the scale to represent larger error on the ground. For example, if a feature on a map is 1 mm from where it should be on a 1:10,000 map, this represents an error of 10 metres on the ground, whereas on a smaller-scale 1:100,000 map this will be an error of 100 metres on the ground.

Paper may warp over time, especially if it has been kept in damp conditions. If a map has been folded, completely flattening it on the scanner or digitiser will be impossible, and the resulting distortion will be reflected in the digital copy. If photocopies or other reproductions are used to capture data, these will have additional error within them. Even without folds, attaching maps to scanners or digitisers, such that they are completely flat, is often not easy. The limitations of the scanner or the digitiser may also cause error or, at least, limit the maximum standards of accuracy.

Digitising will add further sources of error. The process relies on an operator placing the puck or cursor in exactly the right position. Even highly motivated operators may become careless after spending a long day doing this kind of work. Often, the exact location of a feature may not be discernible. If a point is represented by a large dot, it may not be easy to click in exactly the centre of the dot. Thick lines pose similar problems. Dealing with curved lines is always somewhat subjective. These are captured by clicking wherever the line takes a significant change in direction. However, on a smooth curve, such as one found on a railway line, this is a subjective choice. The number of points to use is also subjective. In theory, more points give a more accurate representation, but only within certain limits, and large numbers of points may lead to a large amount of redundancy. As digitising curves relies on subjective judgements, no two operators will digitise the same line in exactly the same way. This is shown in Fig. 3.1 where the two dashed lines represent two different attempts to digitise the same curve. In reality, many more points would be used. However, the underlying problem is that there is no such thing as a true curve in GIS; there are merely straight lines between points. The choice of where to locate these points is subjective.

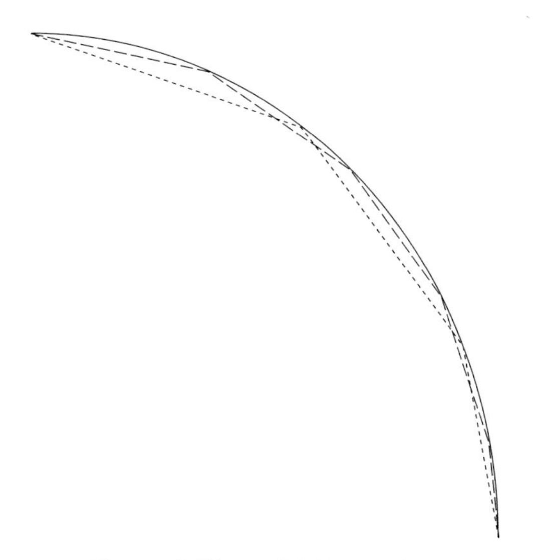

Fig. 3.1 Different digitising producing different lines. This diagram shows two different attempts to digitise the same curve. The curve is shown in the solid line while the two different attempts are shown using dashes. Different choices of where to locate the points that define the curve have resulted in two different versions of the line.

3.3 BRINGING CAPTURED DATA INTO THE REAL WORLD

Whether scanned or digitised, the stages described above leave a layer of spatial data representing a single map sheet whose co-ordinates are usually expressed in inches from the bottom left-hand corner of the scanner or digitiser used to capture it. Making the data fully usable in a GIS may require up to three further stages: geo-referencing to convert from digitiser to real-world co-ordinates; projecting to add a map projection; and joining adjacent map sheets together, probably with the help of *rubber-sheeting*, to create a single layer representing several map sheets.

Geo-referencing is an absolutely fundamental GIS operation. It gives a layer real-world co-ordinates, which means that it can be used to calculate distances, areas and angles in real-world units, and that the layer can be integrated with any other layers that also have real-world co-ordinates. For example, before geo-referencing, we can state that two points are a certain number of inches apart on the page, and that the angle between them is a certain bearing from vertically up on the scanner or the digitiser (which is likely to be close to vertically up the page, but not identical). Once the data have been geo-referenced, we can say that the two points are so many miles or kilometres apart, at a bearing expressed in degrees from north.

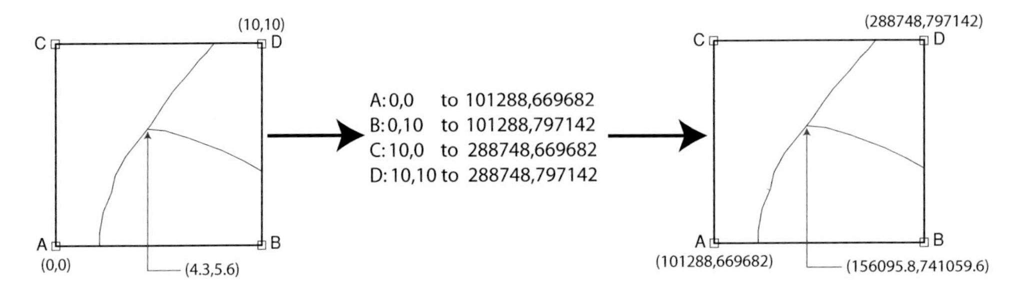

Fig. 3.2 Geo-referencing a layer. The layer on the left has its co-ordinates in digitiser inches. Four control points *A*, *B*, *C* and *D* are chosen and given real world co-ordinates as shown in the centre. These are then used by the software to update all of the co-ordinates on the layer. The co-ordinates of the junction between the two lines are given as an example.

Geo-referencing is done by adding a number of *control points*, sometimes known as *tic points*, to the layer. Four are usually used and, ideally, these are the four corners of the map sheet. These are then given real-world co-ordinates, such as latitude and longitude, or British National Grid co-ordinates. This means that the software knows both the original co-ordinates of the control points and their real-world co-ordinates. It then uses a mathematical transformation to convert all of the co-ordinates on the layer into the real-world co-ordinates. This is shown in Fig. 3.2 where the four corners of the map are used as control points. The user gives the real-world location of these, as shown in the table in the centre, and these updated co-ordinates are used to transform all of the co-ordinates on the layer. The co-ordinates of the junction between the lines are given as an example.

Geo-referencing is clearly an important source of potential errors. If the paper on the map sheet is distorted in any way, or the map has not been placed completely flat on the digitiser or scanner, or there is any inaccuracy in digitising the control points or in measuring their real-world values accurately, these will result in error in the geo-referencing process. This error will remain as a fundamental part of the resulting layer. Most software that allows geo-referencing will give a measure of the error, expressed as a Root Mean Square (RMS) error, usually given in both digitiser units and real-world units. This is an estimate of the error based on comparing the co-ordinates of the control points in digitiser or scanner units against their co-ordinates in real-world units. Using more control points is likely to increase the accuracy of the RMS estimate. An important part of the digitising process is to set a maximum acceptable RMS error. If the RMS error is above this, further attempts need to be made to reduce it by attempting to improve the control points or, if necessary, by re-scanning or digitising the map. For this reason, it is a good idea to check the RMS error before digitising a large map sheet. If the RMS error is within the acceptable threshold, it should still be recorded as part of the documentation that accompanies the dataset (see section 3.6 below). The choice of an acceptable RMS error is subjective and will depend, in part, on the quality of the source. One leading supplier of GIS software recommends not more than 0.004 digitiser inches for good-quality source maps, and perhaps as much as 0.008 digitiser inches for poorer ones (ESRI, 1994).

Most modern maps show a clearly marked grid that can be used to take control points. For example, modern Ordnance Survey 1:25,000 and 1:50,000 sheets show the British National Grid, and intersections from this on or near corners can be used. Older maps may not have this information and this can cause problems. One solution is to find features on the old map that also exist on a modern map. Point features, such as churches, are ideal, but features must be selected with care, as they may have moved over time. These are used as the control points and are digitised using their location from the source map with their real-world co-ordinates being taken from the modern map. This is obviously more error-prone than taking both from the same map, and must be used with care.

Once a layer has been geo-referenced, it is also desirable to give it a map projection. Usually, the layer is first projected onto the same projection used on the source that it was captured from. From here it becomes a relatively trivial job to change projections. Using GIS simplifies the use of projections by making it easy to convert between them. However, all projections have limitations, and it is important to be aware of these. Projections are a complex subject on which extensive literature is available. Dorling and Fairburn (1997) provide an easy introduction, while Bugayevskiy and Snyder (1995), Robinson *et al.* (1995), Roblin (1969) and Steers (1950) provide more detailed descriptions. The intention here is only to give a brief introduction to projections, to explain why they are important.

Projections are necessary because the Earth is a globe, while maps, whether on paper or computer screens, are flat. A point on the globe is expressed using latitude and longitude: the number of degrees, minutes and seconds that the point is from the equator and the Greenwich Meridian respectively. These are sometimes known as *northings* and *eastings* respectively. A location on a map is expressed in x and y co-ordinates: the distance horizontally and vertically from a point that is used as an origin. The process of converting from a round Earth to a flat representation is sometimes compared to peeling an orange and then trying to flatten out the peel. To do so effectively inevitably leads to some distortion and the impact of this distortion needs to be understood. The classic map of the world uses a Mercator projection. It allows the world to be represented on a single rectangular piece of paper. Doing this means that the poles, which are points on the globe, have to be stretched to become lines of the same length as the equator. This means that the scale in an east–west direction becomes increasingly large as we move away from the equator. It is sometimes argued that this projection is favoured in Europe and North America because it stresses places at this latitude at the expense of continents such as Africa.

All projections distort; it is an inevitable consequence of producing a flat representation of a sphere. There are four properties that are affected: angles (direction) between locations, distance between locations, the shape of objects and the areas of objects. Different projections try to preserve one or more of these, but do so at the expense of distorting the others. This distortion tends to become increasingly apparent further away from the projection's central meridian – the equator in a Mercator projection – although other lines are also used.

The British National Grid is a Transverse Mercator projection. It is 'transverse' because the central meridian is at right-angles to the equator. It uses 2°W as the central meridian – a line which runs north from the Isle of Wight, roughly splitting the country in two halves. The projection is scaled such that error is minimised not on the meridian itself but on two parallel lines approximately 200 km east and west of it (Steers, 1950). As the UK is a long, thin country running approximately north–south, this means that errors tend to be minimal. In larger countries such as Australia and America, projections are more problematic. Co-ordinates in the British National Grid use a false origin just south–west of the Scilly Islands. This is the extreme south–west corner of the country and means that co-ordinates can thus be expressed as positive values from this point. Metres are used as the basic units of measurement for both x and y. Therefore, a point at 253425,720621 is just over 250 km east and 720 km north of the false origin, placing it in the southern Highlands of Scotland. All locations on mainland Britain (but not Orkney or Shetland) can be expressed as a six figure reference in this form, as no locations are more than 1,000 km from the false origin. To complicate matters, the Ordnance Survey often gives letter codes such as 'TQ' to the 100 km grid squares. The 100 km grid square at 200 km east and 700 km north is 'NN'; therefore, the co-ordinates above become NN 53425,20621. These references are not helpful in a computer and it is usually preferable to use the full numeric co-ordinate.

Finally, it is often desirable to join adjacent map sheets together to create a single layer. Once the spatial data have been geo-referenced and projected, this is relatively simple. In theory, the co-ordinates at the edges of the sheets should match perfectly; however, in reality, this rarely happens. This makes it necessary to distort one layer slightly to get it to match the adjacent layer. This is a process known as *rubber-sheeting*, as it implies that one sheet is stretched in a variety of directions to get it to fit exactly to the adjacent sheet. This is shown in Fig. 3.3. The basic process usually involves

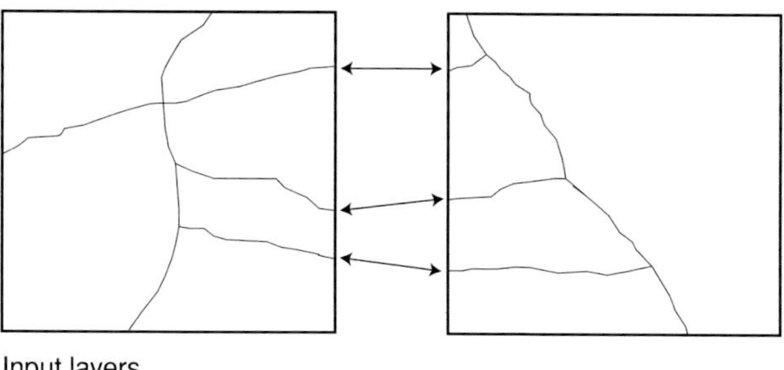

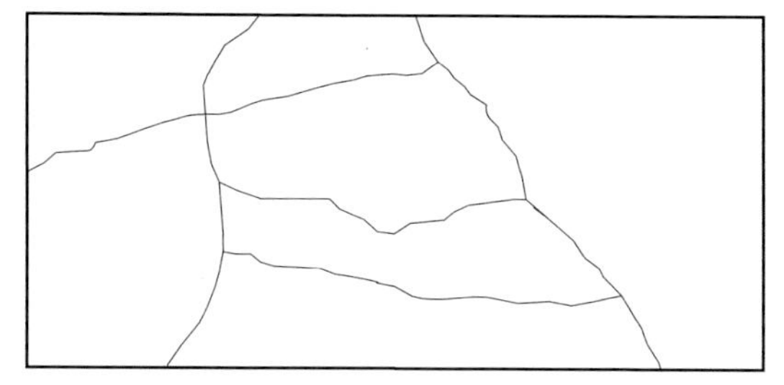

Fig. 3.3 Rubber-sheeting two layers together. In this example, two line layers are to be joined. The lines on the left input layer are held constant and the ones on the right are distorted to ensure a perfect join.

finding lines that run from one sheet to the next. The operator selects lines from the layer to be rubber-sheeted, and indicates which lines from the adjacent layer these are to be joined to. The software then moves the appropriate line to join to the other, but rather than put a sudden kink in the line at the join, the entire sheet is distorted slightly to make the transition smooth. The maximum distortion is at the join, and it becomes less the further away from the join. This is, of course, an additional source of error, and must be done with caution. Usually, a maximum distance that the end of a line will be distorted by is specified, to avoid making this error too large. As long as the original surveys and publishing was accurate, the digitising was performed accurately and the geo-referencing was done properly, the error introduced by rubber-sheeting should be low. With historical maps, achieving these high standards of accuracy is not always possible.

3.4 PRIMARY DATA CAPTURE

Although most historical GIS data will be taken from secondary sources, primary sources should not be overlooked. The two main primary sources of GIS data are from the Global Positioning System (GPS) and remote sensing. GPS enables an operator to capture co-ordinates on the ground to a high degree of accuracy, usually by holding a GPS receiver at that location. Remotely sensed data, usually from satellites – although aeroplanes and the Space Shuttle are also possible platforms – give a raster representation of the Earth's surface that is well suited to use in GIS. As is discussed below, GPS has been used in a number of historical GIS studies; to date, the use of remote sensing is more limited.

The GPS was funded by the US Department of Defense. It is based around a number of satellites with extremely accurate atomic clocks that emit a signal giving their position and the exact time. A receiver on the Earth's surface, picking up a signal from a number of satellites, is then able to calculate its location on the Earth to a high degree of accuracy based on the different lengths of time it has taken the signal to arrive from each satellite. Hoffman-Wellenhof (1997) and Kennedy (2002) provide detailed descriptions of how the GPS works. When it first became available in the early 1990s, the GPS signal was deliberately degraded for civilian users, but even then it measured locations usually to within 100 metres of the actual location. More recently, this *selective availability*, as it was called, has been switched off, and now all users can receive a location to within a few metres of their actual location. The cheapest receivers capable of this only cost around £100. They are hand-held and simple to operate. Improved accuracy, to less than one meter, can be achieved using *Differential GPS* (DGPS), where the location of a roving receiver is compared to the location of a base-station at a known point. This adds considerably to the cost and complexity of GPS equipment.

The GPS allows the historian to capture the locations of points and lines on the ground simply by walking around the features. Lowe (2002) describes doing this to capture the locations of earthworks, foxholes and other historic features from a number of American Civil War battlefields in Spotsylvania National Military Park. Once captured, the co-ordinates can be loaded into a GIS to create vector data. Doing this enabled him to build up a new impression of the battlefield sites based on surviving features that have never been accurately surveyed. In another example, Lilley *et al.* (2005a) describe the use of DGPS to survey the medieval street plan of the town of Winchelsea in East Sussex. They combine their GPS survey with modern Ordnance Survey data and a rental survey dating from 1292, to create a street plan of the town as it would have existed in the thirteenth century.

Remote sensing is often inaccurately called *satellite photography*. It involves a sensor, akin to a digital camera on the satellite or other platform taking an image of the Earth's surface, which is like a complex version of a digital photograph. As such, it makes an excellent source of raster data. Again, it is not the intention here to describe remote sensing in detail, as many good guides are available (see, for example, Jensen, 1996; Lillesand and Kiefer, 2000; Richards, 1993), but a brief discussion will outline what it may have to offer to the historian. Most forms of remotely sensed images simply record the light being reflected back from the Earth's surface, and are thus vulnerable to cloud. Others use radar that can see through cloud. Where satellites are used, images are taken at regular intervals of usually between a few days and a few weeks. Other surveys – for example, the Space Shuttle's LIDAR (Light Detection and Ranging) mission – are one-off surveys that are not repeated.

A basic remotely sensed image subdivides the Earth's surface into pixels, usually squares. Their size varies from a few square metres upwards, although this varies significantly, depending on which type of imagery is used. For each pixel, the image records one or more values that record the amount of light being reflected. This is termed its *spectral response.* A *panchromatic* image records only a single value, usually the amount of light reflected from the Earth's surface to the sensor, and can thus only be visualised using varying amounts of a single colour, usually grey. Most sensors record more information than this, and have multiple values for different wavelengths of light, or *bands.* A simple example would be to have three bands: one recording red light, one green light and one blue light. These can then be combined to produce a *true-colour* image if required. More complex systems record more specific wavelengths of light, some of which may be infrared or ultraviolet, and are set up to respond to specific spectral characteristics of features on the Earth's surface, such as the health of vegetation. Combining the bands in different ways allows images termed *false-colour composites* to be produced. These are so-called because the colours that are used in the display are not necessarily the colour that would be seen from space.

Remotely sensed images are often used in *classified* form. The values recorded for each pixel are used to allocate the pixel to a class, such as dense urban, forest, water, and so on. Depending on the characteristics of the sensor and the classification system used, the classes can be quite detailed. In a *supervised classification,* a sample of pixels from each land-use type is identified, and the software then attempts to identify pixels with a similar spectral response, to allocate these to the same class. In an *unsupervised classification,* the software allocates each pixel to a class based on its spectral response, and the user may then attempt to work out what each class represents on the ground.

A variety of image-processing techniques may be further used to identify patterns from remotely sensed images that may not be apparent to the human eye. These include statistical operations that can be used to identify lines or edges between types of pixels, techniques to smooth out noise from images, and so on (see DeMers, 2002 in addition to the references given above).

To date, there has not been much use made of remotely sensed images by historians. There is the potential to identify features from the past, hidden under the modern landscape, from these images. Archaeologists have made some use of remotely sensed images for these purposes (see, for example, Fowler, 1996; Madry and Crumley, 1990; Riley, 1996; Scallar *et al.*, 1990; Shennan and Donoghue, 1992), and lessons learned from these may be of relevance to some historians.

3.5 LINKING ATTRIBUTE DATA TO SPATIAL DATA

Capturing textual or numeric data is usually done by typing, or scanning and optical character recognition (OCR). Images are usually scanned, and sound and video recorded and digitised in various ways. The detail of doing this is beyond the scope of this book; Bartley *et al.* (2000), Harvey and Press (1996) and Townsend *et al.* (1999) provide descriptions. Frequently it is unnecessary to capture the attribute data, as it is already available in digital form. In this section, we will assume that the researcher has an existing attribute database to which they want to add spatial data to create a GIS database.

There are three ways that attribute data may be linked to spatial data. Firstly, it may be that the attribute data already include spatial data in the form of a pair of co-ordinates for each row. Secondly, there may be a spatial reference, such as a place name, attached to the attribute data that needs to be linked to the spatial data. Thirdly, it may be that there is an existing attribute dataset which needs to have a unique reference added to it to allow it to be linked to the spatial data.

The first situation occurs when a database table that already contains co-ordinates needs to be converted into a GIS layer. The co-ordinates may have been captured as part of the process of creating the attribute database, or may have been added later through, for example, the use of a gazetteer (see Chapter 7 for more detail on this). This situation usually only occurs with point data that can be easily converted into spatial data. Most GIS software can be told which column from the attribute database contains the *x* co-ordinate and which contains the *y*. The software reads these in and creates a layer of spatial data from it, as shown in Fig. 3.4.

A flexible and useful solution to linking spatial and attribute data is where the researcher has a set of attribute data that includes a spatial reference, such as a place name, and a layer of spatial data can be found that includes a similar spatial reference. This often occurs, as many datasets have place names attached. A relational join (see Chapter 2) is used to join the attribute data to the spatial data. Assuming that the place names in both datasets are identical, this presents an effective solution

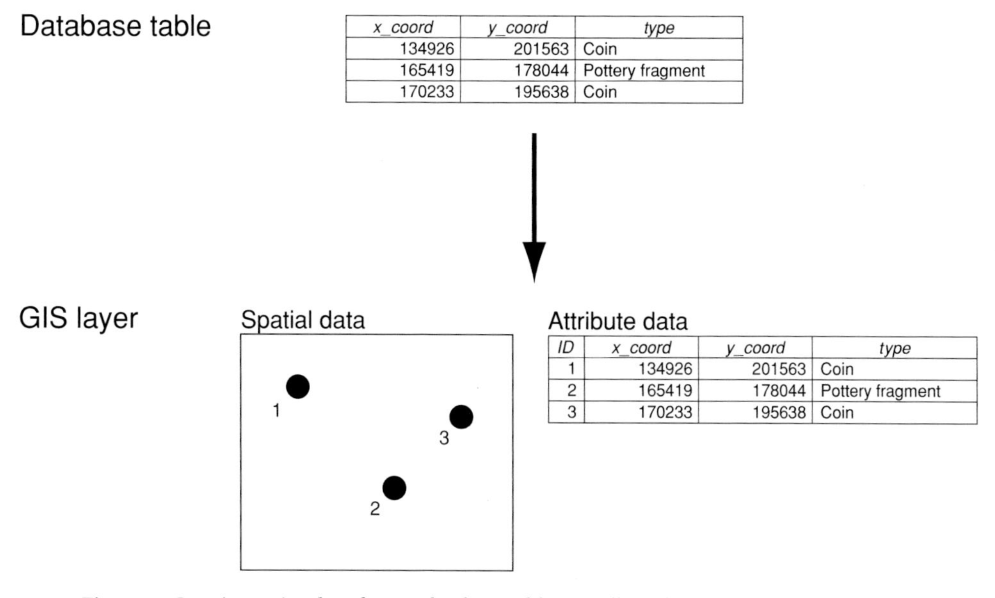

Fig. 3.4 Creating point data from a database table. By telling the software which columns hold the *x* and *y* co-ordinates, the software is able to convert from the database table at the top to the GIS layer shown at the bottom. Note that the co-ordinates usually remain as a part of the attribute data as well as being used to create the spatial data.

to a problem, as is shown in Fig. 3.5. This type of situation often occurs with data for places such as towns or administrative units where spatial data have been created that can be linked to multiple datasets.

One problem that occurs is that spellings of place names are notoriously non-standard and may well change over time or between different sources. This often requires an *ad hoc* solution that simply means that a standardised spelling of the place names needs to be included in both sources to provide a perfect match. A more elegant solution is to build a *gazetteer*, a database table that provides a standard spelling of every place name that occurs in the attribute database. An example of this for the data in Fig. 3.5 is shown in Fig. 3.6. The gazetteer table has two columns: one containing the standardised spelling of the place name as used in the GIS attribute data, and the other containing alternative spellings, some of which may be the same as the standard version. Each possible different spelling requires a new row of attribute data. A three-way relational join is then able to link from the GIS layer to the external database table, taking into account that, for example, 'Compton Verney' in the GIS is spelt 'Compton Varney' in the external database, and that 'Burton Dassett' is hyphenated in the external database. As we can have multiple alternative spellings, this structure can be used to link a single layer of GIS data to multiple rows of attribute

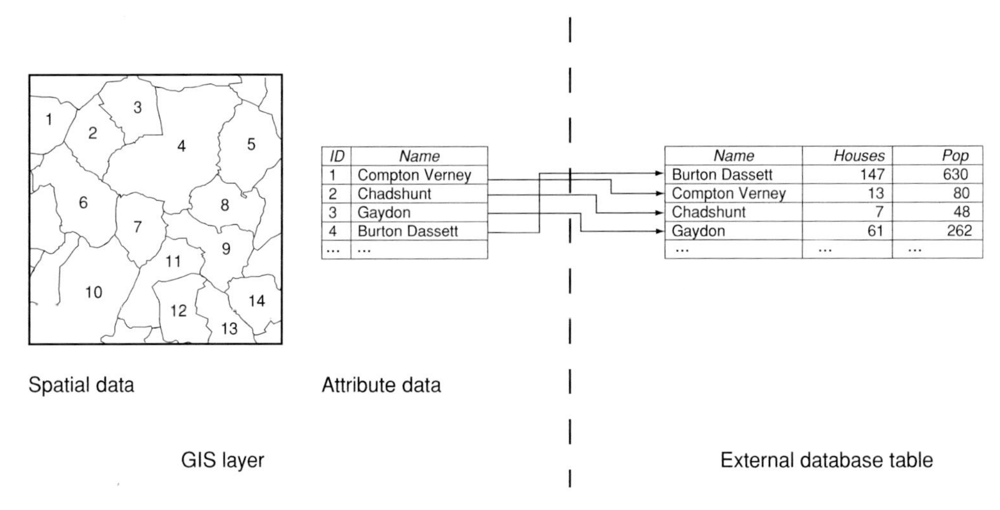

Fig. 3.5 Linking a GIS layer to external attribute data using place names. A relational join on place name between the attribute data and the external database is used to tell us that, for example, polygon 4 is Burton Dassett and has a population of 630.

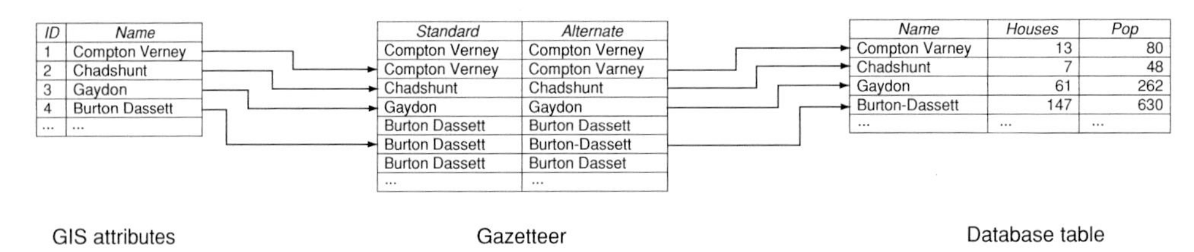

Fig. 3.6 Using a gazetteer to standardise place names. This example takes the same names from the GIS attributes as Fig. 3.5, but in this case there are some differences in the spellings used in the external database table. To handle this, a gazetteer table is used that is able to translate from the standard versions of the place names held in the GIS to the alternate versions held in the external database. Thus, there is enough information that a relational join can join 'Compton Verney' to 'Compton Varney' and 'Burton Dassett' to 'Burton-Dassett'.

data. Thus the gazetteer in Fig. 3.6 could be used to link to three different database tables, each of which has a slightly different spelling of 'Burton Dassett'. More details on the use of gazetteers are given in Chapter 7.

A second difficulty is that the definitions of places may not be identical between the two sources. This is often true when linking data published by administrative units, such as districts or census tracts, to spatial data. Administrative boundaries change frequently, and in some cases the precise definitions of a type of unit may vary between different organisations. Therefore, a layer of spatial data that is intended to represent the attribute data may not, in fact, be completely accurate. Whether it is suitable remains a choice for the researcher.

The final option occurs where both the spatial and attribute data have some sort of name or other identifier, but these are not standard between the two sets. This may occur because of the use of different languages or may be because both datasets have been created with custom identifiers, and their integration was never planned. This is the most labour intensive part of the process, as it often requires the operator to work through the data manually, creating a field in both datasets that has common values and allows a relational join to be created.

A variation may occur when spatial data are being digitised specifically to link to an existing attribute dataset, and where it is easy to add unique ID numbers to the spatial data as they are being digitised, but much more labour intensive to add textual names. In this case, adding a field of unique ID numbers to the attribute data before the data are digitised, and then adding these same ID numbers to the spatial data as the data are being digitised, may result in major time savings.

In all of the above cases, it is important to store the attribute data in the most suitable software package, rather than necessarily importing it into the GIS. Almost all GIS software allows this and, as was discussed in Chapter 2, it greatly enhances the functionality and ease of use of the resulting database. The use of a gazetteer provides powerful relational database functionality that complements the GIS's co-ordinate-based approach well. It is discussed further in Chapter 7.

3.6 ENSURING DATA ARE USABLE: METADATA, DOCUMENTATION, PRESERVATION AND DISSEMINATION

Creating GIS databases always represents a significant time commitment, and often requires the researcher to secure major investment from external funders. To justify these, it is important to consider issues of long-term usability from the outset of a project. Indeed, if external funding is sought, applications are unlikely to be successful if these issues have not been well thought through. To ensure long-term usability, a database must have accompanying metadata and documentation to describe its data. It is also important to have a preservation strategy, to ensure that the dataset remains usable over the long term, in the face of inevitable technological changes. Finally, if external funding is sought to create a database, it is usually a condition of this funding that the data will be disseminated to other users. Developing a strategy to do this is therefore essential.

A fundamental part of creating a dataset is documenting it in such a way that its contents are thoroughly described. This ensures that the person who created the database can remember the sources and design decisions that were taken as part of the construction process. By their nature, GISs can draw on multiple sources and involve many complex judgements during creation. For scholars to use their GIS, it is essential that they can fully recall these decisions, as these may affect the research findings and outputs from the GIS. Beyond this, it enables other potential users to find the dataset in a digital archive or data library, to evaluate whether the dataset is fit for their purpose, to allow scholarly criticism of the database, and to ensure that proper acknowledgement is made to the person or persons who first created it. Although metadata is a form of documentation, in this section a distinction will be drawn between the term *metadata* and the term *documentation*. Metadata are data that describe the content, quality, condition and other characteristics of the data held within a database. They use a database-type structure and tend to follow formats such as Dublin Core (described below). Metadata originated from fields such as information services and library sciences. Documentation, on the other hand, is a far looser concept. It is the use of information to describe the dataset in what is typically a more flexible form than metadata. Historians have a long tradition of using documentation through footnotes and similar techniques. Digital resources, such as historical GIS datasets, can be documented using these techniques to the same rigorous standards as would be used in, for example, transcribing an archival source. Although metadata and documentation come from different fields and may use very different structures, they are not mutually exclusive and are often highly complementary to each other. Metadata provides a summary of the dataset in a highly structured form that enables the user to quickly find the dataset and find out its key features. Documentation provides a far more in-depth examination of the dataset, allowing a critique of its scholarly benefits to be made and enabling the researcher to decide whether it is fit for the purpose that they want to use it for.

A variety of metadata standards are available, of which Dublin Core is perhaps the most widely used (see Dublin Core Metadata Initiative, 2005; Miller and Greenstein, 1997). The use of Dublin Core effectively creates a database table that contains information about the dataset being described. This information includes the title of the dataset, who created it, the subject it is concerned with, a description of the dataset, names of people or organisations who contributed to it, intellectual property right issues, and so on. A full list of fields, or *elements*, used by Dublin Core is given in Table 3.1.

It will be clear from Table 3.1 that Dublin Core was designed to be as broadly applicable as possible, and its criteria are not in any way driven by the needs of

Table 3.1 *A brief description of the elements of the Dublin Core metadata standard*

Element Name	Description
Title	A name given to the dataset
Creator	The person or organisation who created the resource
Subject	The topic that the dataset is concerned with, usually expressed as keywords
Description	A brief description of the resource, which may be in the form of an abstract
Publisher	The entity responsible for making the resource available
Contributor	An individual or organisation responsible for making contributions to the content of the resource
Date	The date when the data were either created or made available
Type	The nature or genre of the dataset, usually expressed as a value from a controlled vocabulary
Format	The software or hardware that are needed to use the dataset
Identifier	An unambiguous reference to the dataset within a given context
Source	The resource that the data were derived from
Language	The language used for the content of the dataset, usually expressed using a standard list of definitions
Relation	A reference to a resource that is related to the dataset
Coverage	The extent or scope of the dataset that usually includes its spatial location and temporal extent
Rights	Information on the intellectual property rights and copyright associate with the dataset

Note that not all elements are compulsory and that there may be more than one value for some elements, such as 'contributor' and 'relation'.
Source: The Dublin Core Metadata Initiative, http://uk.dublincore.org/documents/dces. Viewed 10 June 2007.

historians or of GIS data. It is particularly weak at handling geographical and temporal metadata using only a single broadly defined element: *coverage*. Other metadata standards have been developed explicitly to handle GIS data, such as those devised by the Federal Geographic Data Committee (FGDC) in the US (FGDC, 2005; Guptil, 1999; Intra-Governmental Group on Geographic Information, 2004). Even these may not be appropriate for the type of data within a historical GIS database. As a result, we recommend that researchers considering how to document a database think carefully about how to do this. Their decisions need to be based in part on metadata standards but also the usual approaches to documentation common in their discipline. Using non-standard metadata creates its own problems. There is an increasing trend to allow multiple metadata catalogues to be queried simultaneously over the internet, bringing multiple catalogues together through a single portal. Doing this requires metadata to be in standardised forms; therefore, using non-standard metadata may not be compatible with the requirement that metadata should allow potential users to find the resource.

A final issue concerning metadata and documentation is at what level it should be applied. Clearly, it is important to document the dataset as a whole, but if a dataset brings together elements from multiple sources with, perhaps, varying scales and standards of accuracy, it may be desirable or even essential to document each individual feature. This needs careful consideration, as documentation at this level is time-consuming to create, but a lack of it may fundamentally limit the usefulness of the database.

A long-term preservation strategy is required if the data are to remain usable over time. Software evolves rapidly with proprietary file formats – those owned by a single company – being particularly vulnerable to change. New versions of software often use updated file formats; therefore, *backward compatibility* – the ability to read the older formats – may not last long. Worse still, proprietary file formats may disappear altogether if the company that owns them goes out of business. Non-proprietary formats, those that are an industry standard, such as ASCII text (American Standard Code for Information Interchange), HTML (Hypertext Mark-Up Language), XML (Extensible Mark-Up Language) and TIFF (Tagged Image File Format), are less vulnerable to change, but do get updated and are often less standard than may be expected. Hardware and the media on which data are stored also become out-dated rapidly, and many once-standard pieces of equipment such as punch-cards, tape readers and 5¼″ floppy disk drives are now all but unreadable.

To attempt to ease the preservation issue from a software perspective, a limited number of file formats have become accepted as standards that are likely to exist in the long term (or at least longer term than most other formats). Originally, this was mainly ASCII text; however, increasingly, other formats, such as TIFFs for images, are becoming accepted. There is, however, no GIS file format that is suitable for this. ESRI's proprietary shapefile format has become an accepted standard simply because it is probably the most widely used; however, it is far from clear that this has a long-term future. Geography Mark-up Language (GML), a text-based extension of XML, has been proposed as a suitable standard, but has not been widely adopted. Although this is far from perfect, it is better than the situation for hardware where we can have no confidence of long-term viability of devices. The 3½″ floppy disc is currently becoming obsolete, and CD-ROMs and DVDs will eventually go the same way.

The combined effect of both software and hardware obsolescence, plus the lack of an accepted and widely used, standard, non-proprietary file format means that even simple GIS databases are vulnerable to becoming unusable. For complex databases that involve unusual database architectures or custom-written software, this is even more serious. There are no strategies currently available that guarantee long-term preservation. However, there are many approaches that guarantee that the data *will* be lost. The most effective of these is for there to be a single copy of the data on a PC or server of an individual academic or organisation. This will almost inevitably result in the data being lost.

Dissemination means making data available to other users. This involves making it known that the data are available, providing access to the metadata and documentation, providing the data themselves in a format and on a media that the person requesting the data can use, and handling requests and issues from these people. Although internet dissemination of digital data may seem easy, as will be discussed in Chapter 7, it is often more complex and time-consuming than is realised from the outset. Handling user queries is a particularly time-consuming process, the costs of which are frequently underestimated.

Effective preservation and dissemination of data is therefore a difficult and specialised task that will frequently be beyond the scope of individual academics or research projects. For this reason, the best preservation and dissemination strategy is usually to give the data to an appropriate data library or digital archive that has the specialist skills required. Examples of such organisations include the Arts and Humanities Data Service (AHDS)[1] and the UK Data Archive (UKDA)[2] in the UK, and the Alexandria Digital Library[3] and California Digital Library[4] in the US.

3.7 CONCLUSIONS

Two brief examples show both the advantages and costs of creating historical GIS databases. Siebert (2000) provides a good example of the issues associated with his construction of a historical GIS database for Tokyo. His database brings together a wide variety of information on the city, including data on the physical landscape, administrative boundaries, data from population and economic censuses, information on commercial and industrial activities, information on the growth of the road and rail networks, and information on land ownership. For most of these, he is incorporating data on change from the mid-nineteenth century to the present. This is a hugely time-consuming operation and he is very honest about the difficulties

[1] See www.ahds.ac.uk. Viewed 10 June 2007. [2] See www.data-archive.ac.uk. Viewed 10 June 2007.
[3] See www.alexandria.ucsb.edu. Viewed 10 June 2007. [4] See www.cdlib.org. Viewed 10 June 2007.

that he has faced in constructing this system. At the time his article was written, he had been working on this database for a number of years and still had a considerable way to go to complete the database.

Gregory *et al.* (2002) provide another example with the construction of the Great Britain Historical GIS. This has a more specific theme than Siebert's database, in that they are primarily interested in changing administrative boundaries and the data that were published using these boundaries, such as the census, vital registration and Poor Law statistics. Nevertheless, to build a system that covered change over time from the mid-nineteenth century to the 1970s, they had to integrate information from a variety of series of maps at different scales, some of which were only available in the Public Record Office (now the National Archive). They also had to add information on when boundaries changed, taken from a variety of textual sources, and also had to integrate a large amount of statistical information in an attribute database, which they did using place-name gazetteers. Covering England and Wales alone took several years and over £500,000 of funding. By the end of the process, they had created an integrated system containing a database of changing boundaries over time, and a database of censuses and other major historical sources over time. Within the GIS, all of the statistical data can be linked to an administrative unit. This is not the case on paper or in any other form of database technology. In this way, the GIS reunites data that were separated in the paper-based publication process. As they do this for change over time, they also have an integrated database of the changing society of England and Wales, created in a way that would not have been possible without GIS. For more detail on this study, see Chapters 6 and 9.

Building historical GIS resources is a time-consuming and expensive undertaking that gives long lag times before results start to become achievable. Before beginning to build a GIS database, it is necessary to carefully consider these costs and to determine whether the end product justifies them. If the end product is simply a map or series of maps, it is unlikely that these will justify the construction costs, and it may be that other approaches such as digital cartography are more appropriate. On a more positive note, however, a well-constructed and properly documented GIS database provides a research resource structured in ways that no other approach can manage, and often allows integrating data from a wide variety of disparate sources and different dates. These have the potential not only to assist in answering the specific research questions that the researcher may be interested in, but also to provide resources for scholars for many years to come.

4

Basic approaches to handling data in a historical GIS

4.1 INTRODUCTION

Chapter 2 described how GIS represents the Earth's surface using vector and raster models that contain a combination of spatial and attribute data. This chapter explores the basic tools that GIS offers to exploit the data held within a historical GIS, and the limitations of them. In many ways, these tools are surprisingly limited; the one thing they have in common is that they provide ways of handling data in a manner that explicitly exploits the spatial characteristics of the data. This is why GIS is an effective instrument for geographical scholarship. The tools described here form the basic building blocks of the more advanced functionality and methodologies described in subsequent chapters.

GIS software provides three basic sets of tools. The first tool is querying the database using spatial, attribute or topological information. This allows data to be retrieved from the database using either the spatial or attribute data to allow geographical questions to be answered. The second tool is manipulating the spatial component of the data to allow new data to be created. This can involve converting a polygon layer to points defined by the *centroid* of the polygon or converting points to polygons using Thiessen polygons. Polygons can also be aggregated together based on the values of their attributes – a process known as *dissolving*. The third tool allows data to be integrated using their location, either informally by displaying two or more layers on the same map, or formally using a geometric intersection.

The ability to manipulate data using these approaches clearly allows sources to be used in new ways. However, it also means that the user is making increased demands on the quality of the data, and in particular on its spatial component. Historians are used to working with data whose quality is limited, but are less used to the problems associated with the limitations of spatial data and, in particular, how GIS may pose demands on these data. This chapter thus also explores issues associated with data quality within historical GIS.

Querying, manipulating the spatial component of the data, and integrating data through location provide the basic building blocks to much of what follows in this book. They are, however, technological fixes that must be used with caution and within the limitations of the source data. When using them, the historian must be aware not only of the limitations of the source as they would conventionally think of them, but also of the limitations from a GIS perspective. As long as these precautions are taken, the set of tools in this chapter provides the historian with the power to explore, manage and integrate historical resources in ways that no other approach allows.

4.2 EXPLORING DATA THROUGH QUERYING

Querying involves retrieving data from a database in a form that makes it understandable to the user. It is an operation that is fundamental to all databases, but within GIS it becomes both more complex and more powerful because of the combination of attribute and spatial data.

The conventional form of query in a database is the attribute query. These can take the form of for example, 'select all of the places where unemployment is greater than 10 per cent' or 'select towns whose type equals "BOROUGH"'. These queries are often implemented using a query language called Structured Query Language (SQL), although this is often hidden from the researcher through the use of a graphical user interface (see Harvey and Press, 1996 for historical examples). In a conventional database, queries of this sort will produce a list of the rows of data that satisfy the criteria specified. A GIS database will also do this, but, because each row in the GIS is linked to an item of spatial data, GIS is also capable of showing the user *where* the features selected are found.

Figure 4.1 shows an example of an attribute query. There is a polygon layer representing administrative units whose attribute data include information on unemployment rates. In the query, all areas with rates of over 10 per cent have been selected. These are shown with ID numbers and attribute data highlighted in bold. As the diagram shows, by querying the attributes, the user is able to find information about the spatial patterns. In Fig. 4.1, it is clear that the higher unemployment rates are concentrated in the smaller administrative units, which are almost certainly urban areas. As well as spotting the broad spatial patterns this form of query can also reveal exceptions to them – for example, polygons 5 and 7 are not selected by the

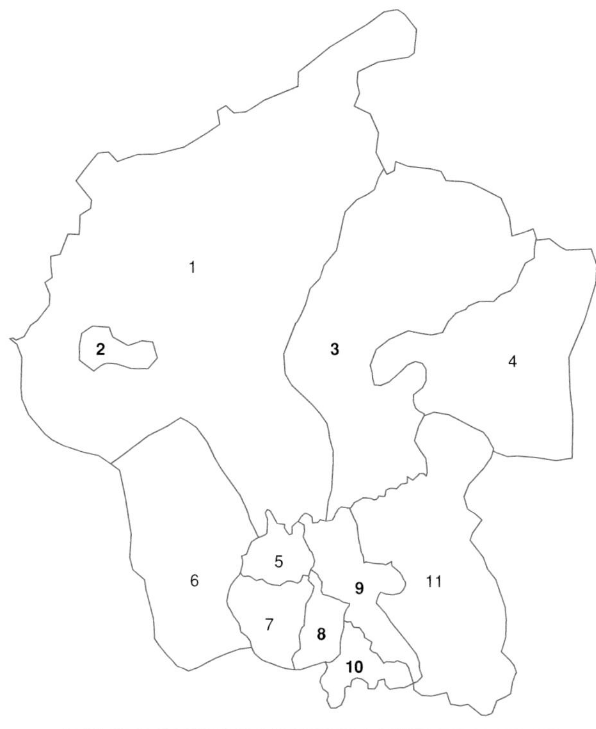

ID	*Name*	*Unemp*	*Pop*	*U_rate*
1	Henny	103	2,154	4.8
2	**Slimbridge**	**716**	**5,821**	**12.3**
3	**Dale**	**120**	**1,017**	**11.8**
4	Maldon	55	1,081	5.1
5	Newland	272	4,247	6.4
6	Amprey	87	2,988	2.9
7	Warterstock	511	5,211	9.8
8	**St. James**	**781**	**5,005**	**15.6**
9	**St. Peter**	**1,174**	**6,488**	**18.1**
10	**St. Paul**	**823**	**5,586**	**14.9**
11	St. Giles	106	1,047	4.1

Fig. 4.1 Attribute querying of polygon data. The polygon layer represents administrative units whose attribute data includes numbers of unemployed people (unemp), total population (pop) and the unemployment rate as a percentage (u_rate). The query has selected all items of data with an unemployment rate of over 10 per cent. These are represented with bold polygon identifiers and attributes.

query, unlike the other apparently urban areas, while polygon 3 is selected, unlike other apparently rural polygons. Through attribute querying, the researcher is able to explore the spatial patterns within the data, and can begin to consider the processes that may be causing them.

Attribute querying underlies much of the mapping performed by GIS. Visualisation will be dealt with in detail in Chapter 5; however, taking the database given in

Fig. 4.1, it is clear that producing a choropleth map of these data, where rows of data with unemployment rates of less than 5% are shaded using a light shading, those with values from 5 to 10% are shaded using a darker shading, and so on, can be done using attribute querying. Firstly, the polygons with values of less than 5% are selected using an attribute query. These are shaded in one shade. Then, values between 5 and 10% are selected and shaded in a darker colour, and so on.

As well as querying by attribute, GIS data can also be queried using the spatial data. This is usually done by drawing the layer on-screen and selecting individual features using the cursor. It might equally involve selecting all the features within a box or polygon drawn by the user. The attributes of the selected features are then listed so that they can be explored.

To go back to Fig. 4.1, a spatial query might involve clicking on polygon 5, in which it returns that it is called 'Newland' and has an unemployment rate of 6.4 per cent. Equally, a box could be defined around the small polygons near the bottom of the figure and all of their names and values would be listed.

Spatial querying can be done with all forms of data, but is particularly useful with unconventional types, including images, movies, sound files and even web pages with hyperlinks to other resources. An example of this is shown in Fig. 4.2, where there is a point layer representing sites of interest within a town. These include photographs of buildings and scans of archaeological artefacts. By clicking on a point, the user is able to examine the images (or sounds or movies) or to look at web pages about the location and follow hyperlinks either to other pages in the database or to external websites that provide more information. This is the approach followed by the Sydney TimeMap project (Wilson, 2001; see also later in this chapter) at the Museum of Sydney, where it has proved to be highly successful in structuring a database of artefacts about Sydney in a manner that is easily accessibly to the general public.

Spatial and attribute querying are fundamental to GIS. Less commonly used, but equally useful in certain contexts, is topological querying. As was introduced in Chapter 2, topology provides information about how features connect to each other. It has two main uses in GIS: firstly, it gives a structure to polygon data so that for each polygon, the computer stores information about the lines that make up its boundaries, and for each boundary it stores which polygon lies on its left- and right-hand sides; secondly, it is used in line data to create networks, such as information about which lines connect at which junctions.

Topological querying uses this information to answer questions about how features within the database connect to each other. A good example of a topological query would be selecting a polygon representing an administrative unit to find out which administrative units are adjacent to it. This may provide useful information in its

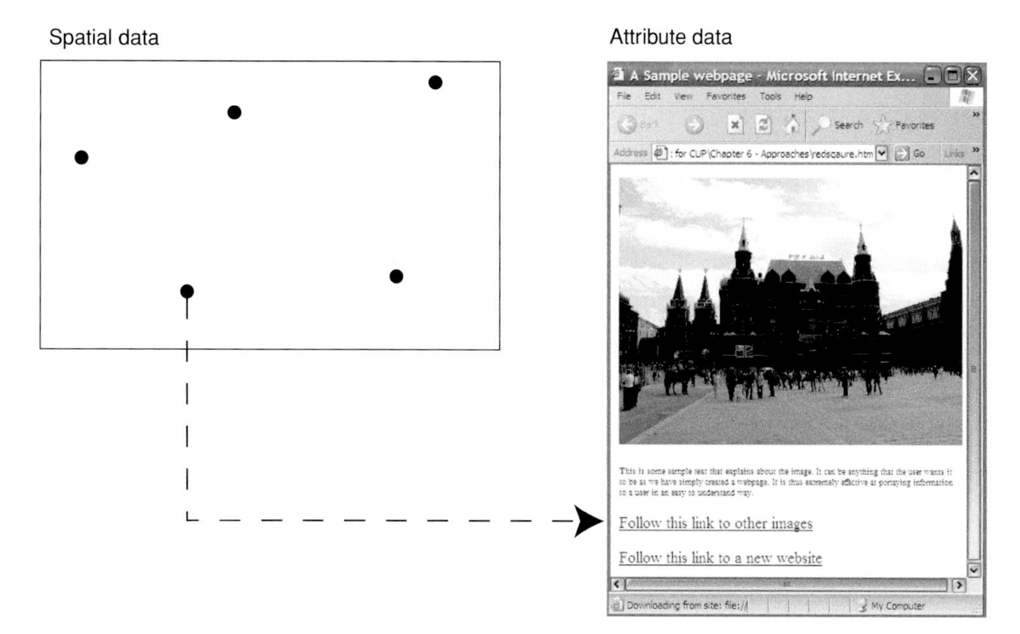

Fig. 4.2 Spatial querying of qualitative information. Clicking on a point in the spatial data returns attribute data in the form of a web page. In this example, the page includes an image of what is at the location plus some text and hyperlinks to other images in the database and external websites.

own right, but is also fundamental to statistical techniques that modify the value of an area according to the values of its neighbours. Techniques that do this are introduced in Chapter 8. Many types of GIS software do not explicitly provide functionality to perform queries of this sort, but sufficient information is available within the data files to allow these to be performed.

Topological querying is also fundamental to networks where it is used to query which lines join to other lines. This is essential functionality, as it allows for queries such as 'how do I get from point *A* to point *B*?' This can be made more complex by adding attribute data representing, for example, the transport cost of travelling along a line segment. This cost may be in distance, time, fuel used or any other cost that is incorporated into the attributes. With this information available, topological querying then allows queries such as 'how much will it cost to get from *A* to *B* using this route?' or 'what is the cheapest way of getting from *A* to *B*?'

GIS therefore allows spatial information to be incorporated into its querying, to ask *where* something is found using attribute queries, *what* is at a location using spatial queries, or how locations *connect* using topological queries. This means that as a form of database, GIS provides the user with an approach that stresses the importance of location. A key point is that the presence of spatial data allows the user to interact with the GIS database through a map-based interface. As such, the use of GIS fundamentally redefines the role of the map in research. As soon as the database is created, the map becomes a way into the data, and the researcher will constantly interact with the map as part of the process of exploring and analysing the data that he or she has.

4.3 CENTROIDS, THIESSEN POLYGONS AND DISSOLVING: TOOLS FOR MANIPULATING A LAYER OF DATA

GIS provides a number of tools for manipulating the spatial data in a single layer to produce new data. It is sometimes desirable to convert from polygon data to points. This requires using the polygon's centroid. It may also be desirable to convert point data to polygons – for example, to estimate the sphere of influence around towns. This is easily (although perhaps not realistically) done using Thiessen polygons. Finally, it may be desirable to simplify a complex layer of polygons to a more generalised layer, by aggregating polygons with the same attribute value. This might happen if, for example, a user has a polygon layer representing fields with crop type as an attribute. If they are only interested in crop type, the user may want to aggregate adjacent fields with the same crop type together, to create a simpler layer. This is called *dissolving.*

The *centroid* is the point at the geographical centre of a polygon. Its location is easily calculated by GIS software. Once this has been calculated, conversion of a polygon layer to points is easy. The polygons' attribute data will usually be allocated to the points in full.

There are a number of reasons why it may be desirable to use points to represent polygon data. A conceptual reason is that it is seen as more realistic to represent a population surface using sample points that are assumed to give information about the underlying population, rather than polygons for which we have no indication of how the data are located within each polygon. This concept was used by Bracken and Martin (1989; see also Chapter 8) to redistribute population data from the British census. Another reason is that it may be desirable to measure the distance between polygons. One way of doing this is to use the centroids of the polygons. Finally, it may be desirable to aggregate polygons based on whether they intersect with larger polygons. An easy way of doing this is to aggregate them according to which polygon their centroid lies within. Martin *et al.* (2002) use this in their Linking Censuses through Time project, to aggregate data from the 1991 census, published at Enumeration District (ED) level, to approximate to 1981 wards. As there were around ten EDs to every ward, they claim that this is a satisfactory way of making the two

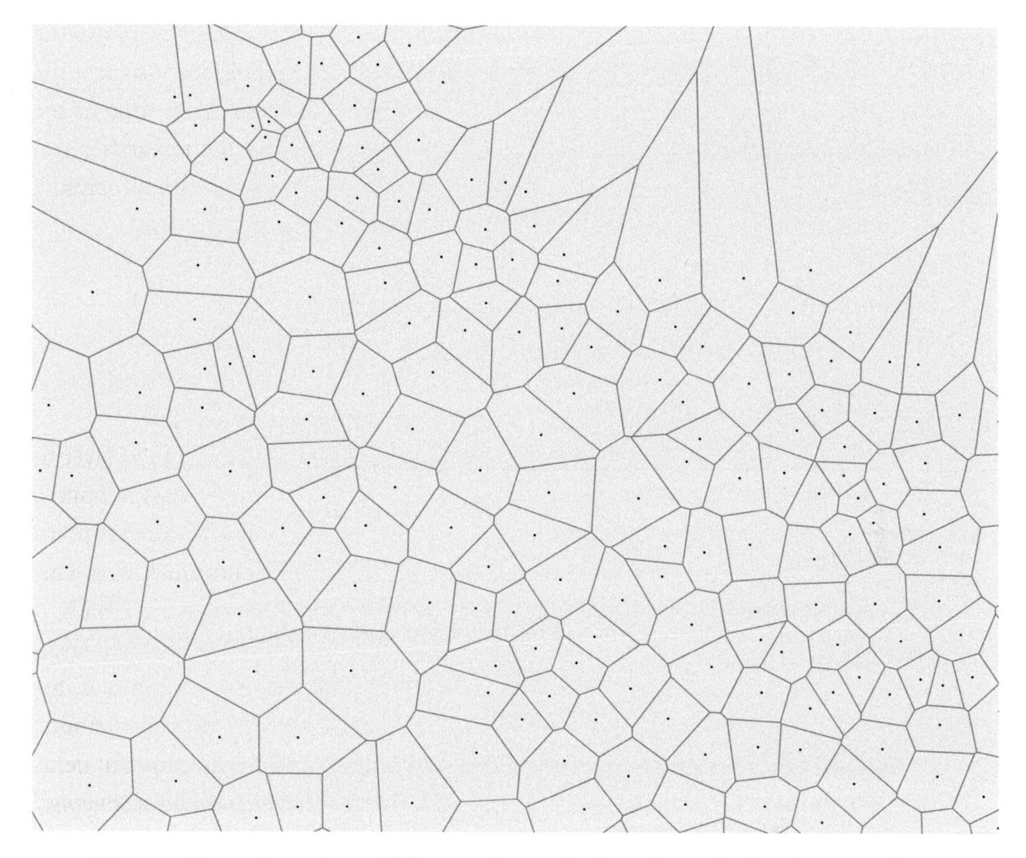

Fig. 4.3 Converting points to Thiessen polygons. An input layer of points is converted into an output polygon layer whose boundaries are equidistant from a pair of points. The Thiessen polygons and the points that created them are both shown.

datasets compatible. As described in more detail below, Gregory (2007) also uses centroids to allocate infant mortality data, published for Registration Districts, to buffers defined by distance from London.

Obviously, the effectiveness of a centroid at representing a polygon is something that the user needs to consider, particularly as the geographical centroid that GIS can calculate may not be the same as the population-weighted centroid – the centre of the population within an administrative unit, whose location is usually unknown. Converting the opposite way – from a point to a polygon – is more difficult, and may be less effective. The easiest way to do this is through the use of Thiessen polygons, also known as Voronoi polygons. These are created from a point layer by putting boundaries that separate each point from its neighbours. The boundaries are located such that they are equidistant to the two points. This is shown in Fig. 4.3. The attributes from the points are given to the newly generated Thiessen polygons.

A A B B B
A C B B B
A A

Input layer

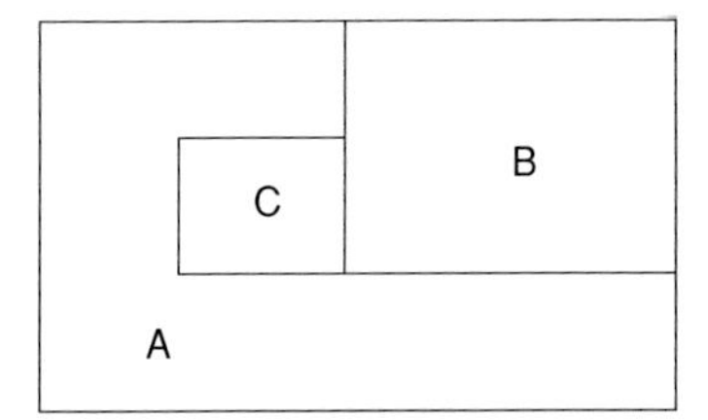

Output layer

Fig. 4.4 Aggregating polygons with a dissolve. Boundaries between polygons with the same attribute values are removed resulting in aggregation. In this case the attributes are simply *A*, *B* or *C*.

Walford (2005) evaluates the use of Thiessen polygons in his attempt to recreate 1971 and 1981 English and Welsh ED boundaries that have never been mapped. He used the centroids of the 1981 EDs, for which locations are available, to create Thiessen polygons. He found these to be unsatisfactory because they have an unnatural appearance and are not suitable for spatial analysis due to their inaccuracy. Instead, he opts for a hybrid approach of using other boundaries that approximate to the 1981 and 1971 ED boundaries, such as 1991 EDs and 1981 wards, but uses the Thiessen polygon approach when none of these are appropriate.

The final technique in this section, *dissolving*, is useful whenever there is a need to aggregate polygons based on their attributes. Boundaries between polygons with the same attribute value are removed to create larger polygons. The basics of this are shown in Fig. 4.4 where the polygons might represent fields and the attribute values represent crop types. Another example might be to aggregate from a low-level administrative unit, such as parishes, to a higher-level one, such as districts. In this case, adjacent polygons with identical district names would be dissolved. There are several options for handling other attribute data that the input layer may have. Take, for example, a parish-level layer for which the data available are parish name, district name, total population, unemployment rate and the class of industry in which most employment is concentrated. These are to be dissolved up to district level, using district name. The attributes for the district-level polygons would obviously include district name. Parish name would be discarded as irrelevant. Total population can be aggregated to give total population at district level which may provide useful new information. The last two variables, however, are rate and class, and these cannot be accurately aggregated. It may be that a weighted average of the rates may be suitable, but equally it may be that this information cannot be used.

Pearson and Collier (1998 and 2002; see also Chapter 2) use dissolves in their work on the tithe survey. This surveyed the country at field level, recording information about the owners and occupiers of each field, the crop grown, the taxable value of the field and a variety of other variables. They use dissolve operations to aggregate several of these variables to higher levels, such as crop type and land owner, before they perform much of their analyses.

4.4 OVERLAY, BUFFERING AND MAP ALGEBRA: INTEGRATING DATA THROUGH LOCATION

The third major set of tools offered by GIS is the ability to integrate data from disparate sources using location. This may be done either informally by superimposing two or more layers on the screen, or formally through an overlay operation which may be complemented through the use of buffering. In either case, this integration has immense potential to bring together disparate historical sources; however, at the same time, it poses new demands on the quality of the data, as error and inaccuracy in spatial data will rapidly spread as more data are integrated. This will be returned to in the next section.

Informal data integration simply involves superimposing two or more layers of data on top of each other on the screen. Doing this uses not just GIS's mapping abilities, but also the ability to change projections and scales. Therefore, given two paper maps with two different projections and at slightly different scales – for example, 1:50,000 and one-inch-to-the-mile (1:63,360), it would be very difficult, inaccurate and labour intensive to integrate these using manual techniques, such as tracing paper. In GIS, however, it is a trivial and accurate task. Simply examining the maps on screen gives visual integration of the information.

A good example of the utility of this approach is described by returning to the Sydney TimeMap project introduced above (Wilson, 2001). This project has the images of many artefacts associated with Sydney's history and development stored as point layers in a database, and aims to use a map-based interface to allow the public to access this database in the Museum of Sydney. On its own, this layer of points is largely meaningless. It requires context in the form of a background map of Sydney. A simple way of doing this is to use a raster scan of a map as a *back-cloth* to the point data. This provides context, as it allows the user to see where each feature is located on the ground. Here, however, another problem arises. As the artefacts are from a variety of dates, clearly a single map – for example, of modern Sydney – is only of limited use. Again, GIS's ability to integrate data becomes useful. The project scanned maps from a variety of dates – from very early maps to the present – and allows the user to turn these on and off as required. As well as providing context to the point data, these maps provide valuable insights into the development of Sydney in their own right. By having them geo-referenced and held within a single GIS database, users are now able to explore the development of Sydney over time, using different maps. This could also be used to provide information about the development of the cartography of the city. In addition, other data, such as areal photographs and satellite images, can be added to compare the city as depicted cartographically with 'bird's eye views'. Other vector data could also be included as required.

Integrating maps in this way also poses additional demands on the source. In theory, a street or other feature that has not moved should be in exactly the same place on every map. In reality, this is unlikely, due to errors and distortions from the original surveys, the map sheets and the digitising or scanning process (see Chapter 3). Where this occurs, it leads to a question of whether inaccurate sheets should be distorted to make them more accurate, or left as they are to preserve the integrity of the source. In reality, preserving the source is likely to be the better option; however, if there are major incompatibilities then there may be a case for distorting the source to allow improved understanding to be gained from it.

The Sydney TimeMap project used custom-written software to manage the data, partly to simplify the user interface and partly to allow time to be handled (see Chapter 6). All conventional GIS software packages provide the functionality to allow this to be done. This involves the ability to change projections if required, and the ability to turn layers on and off and to alter the order in which they are drawn. One source of complexity when multiple layers of raster data are used is that the top layer will cover the features below it. This can be handled in some packages by making the data in one or more layers translucent. An alternative approach is to create vector data of important features, such as the coastline and major roads, from one layer and only superimpose these on other layers.

Formal forms of integration lead to two or more layers being physically integrated. This involves the spatial and attribute data on two input layers being merged, to create a new output layer. In GIS, this is done using an overlay operation (Flowerdew, 1991). In theory, all types of spatial data can be overlaid to produce new layers. Figure 4.5

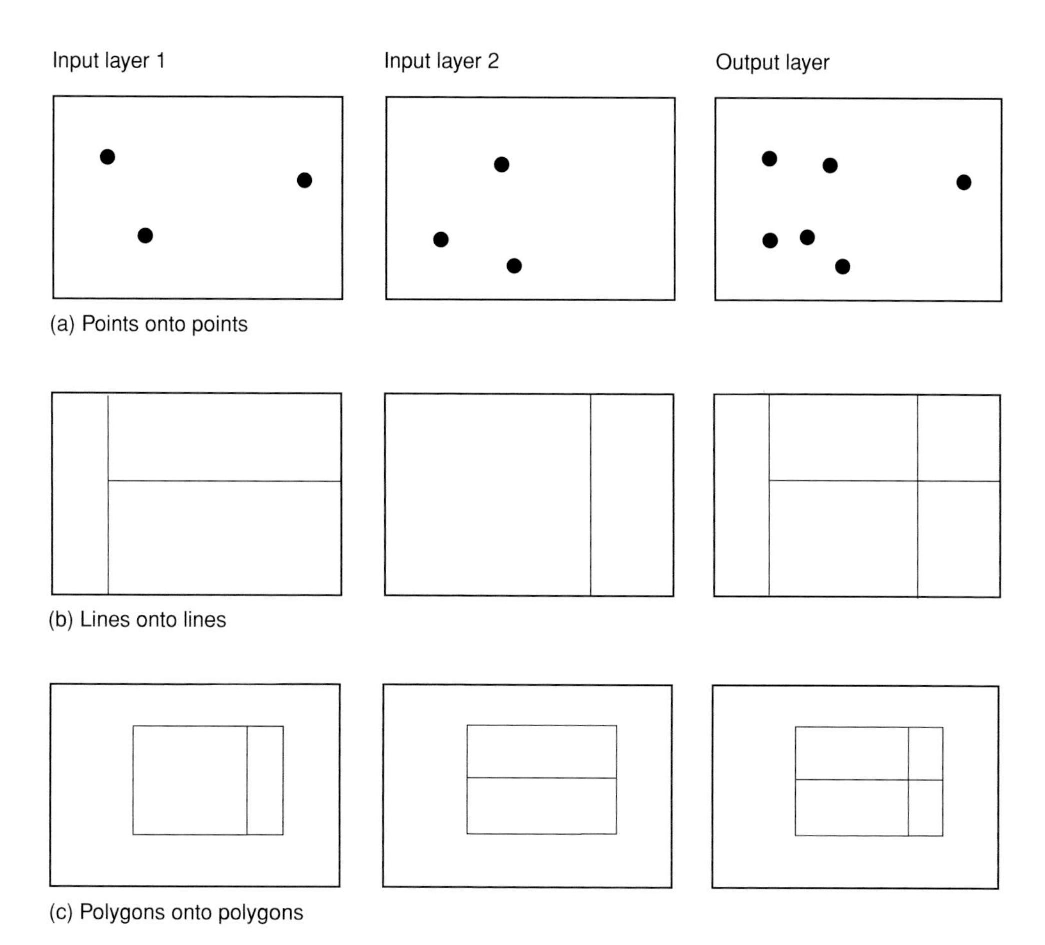

Fig. 4.5 Overlaying vector data to produce new layers. In each of these examples the spatial data on the output layer is derived from the geometric intersection of the input layers.

shows examples of how spatial data are merged by vector overlay. In Fig. 4.5a, two point layers have been overlaid to create a new point layer, Fig. 4.5b uses two line layers, and Fig. 4.5c two polygon layers. In all cases, the output layer is the geometric intersection of the two input layers.

The real power of overlay is that it combines not only the spatial data, but also the attribute data. An example of this is shown in Fig. 4.6, where a point layer representing churches, and a polygon layer representing parishes have been overlaid. The resulting output layer is still a point layer with the locations of the churches on it, but the output layer's attributes have added the parishes' attributes to those from the churches. Figure 4.7 shows what happens when two polygon layers are overlaid.

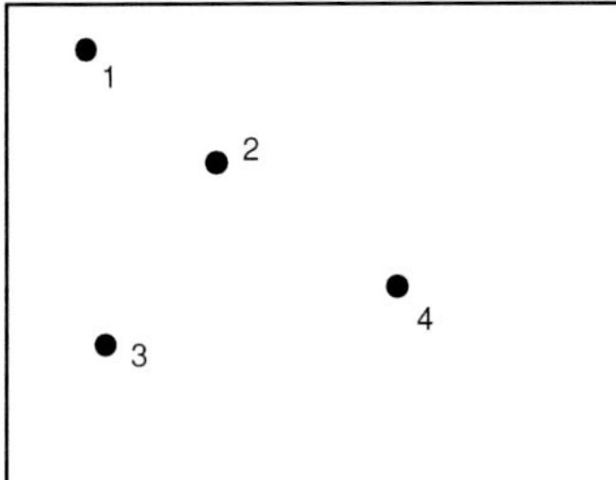

ID	*Name*	*Attendance*
1	St. Stevens	412
2	St. Michaels	396
3	St. Johns	917
4	St. Peters	755

Input 2

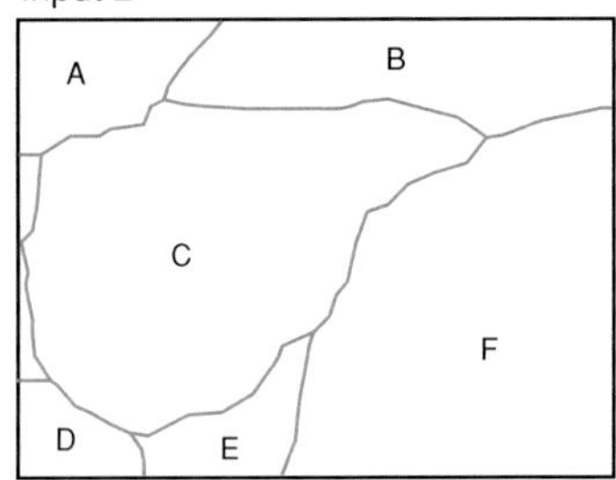

ID	*Parish*	*Population*
A	Irnham	4,362
B	Kencott	8,600
C	Rampton	2,164
D	Durston	2,820
E	Bamford	1,898
F	Sheering	7,311

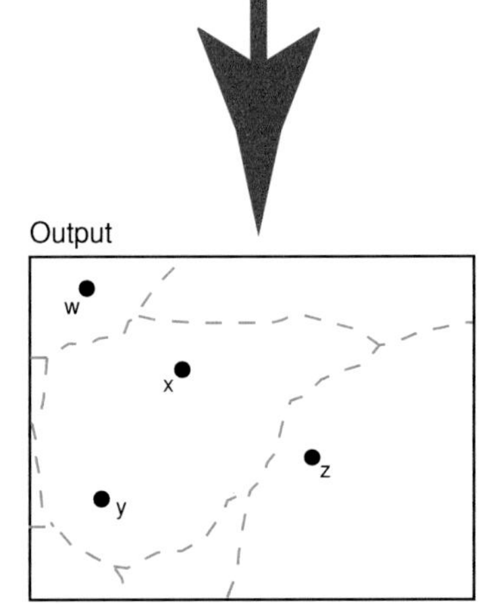

ID	*ID1*	*Name*	*Attendance*	*ID2*	*Parish*	*Population*
w	1	St. Stevens	412	A	Irnham	4,362
x	2	St. Michaels	396	C	Rampton	2,164
y	3	St. Johns	917	C	Rampton	2,164
z	4	St. Peters	755	F	Sheering	7,311

Fig. 4.6 Overlaying points and polygons. In this diagram, a point layer representing churches is overlaid with a polygon layer representing parishes. The output layer is a point layer that has the combined attributes of the two input layers. The outlines of polygons appear on the output layer for diagrammatic purposes only.

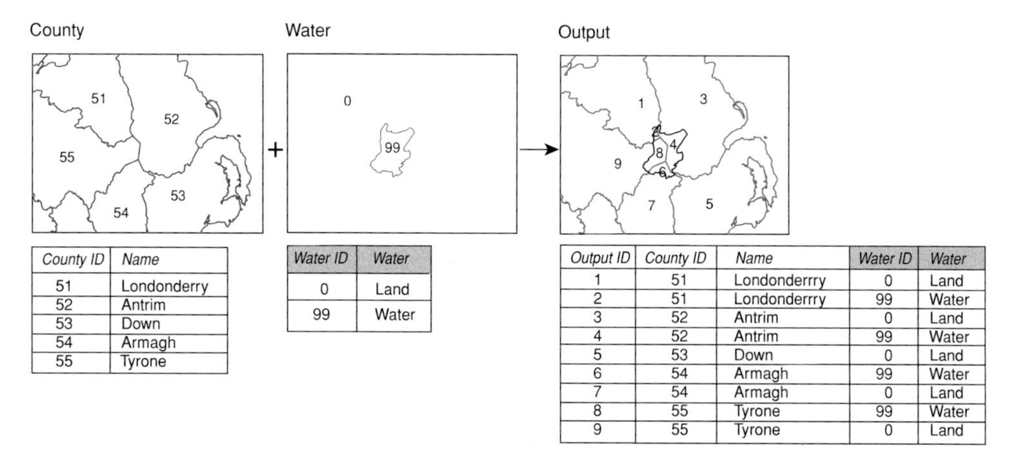

County ID	Name
51	Londonderry
52	Antrim
53	Down
54	Armagh
55	Tyrone

Water ID	Water
0	Land
99	Water

Output ID	County ID	Name	Water ID	Water
1	51	Londonderrry	0	Land
2	51	Londonderrry	99	Water
3	52	Antrim	0	Land
4	52	Antrim	99	Water
5	53	Down	0	Land
6	54	Armagh	99	Water
7	54	Armagh	0	Land
8	55	Tyrone	99	Water
9	55	Tyrone	0	Land

Fig. 4.7 Overlaying two polygon coverages. The input layers representing Northern Ireland counties and inland water respectively are overlaid to merge both the spatial and attribute data.

In this case, the input layers consist of counties in Northern Ireland and inland water for the same area. By overlaying the two, both the spatial and attribute data are merged, such that on the output layer it is possible to see which parts of each county are covered by water. Most GIS software will automatically calculate the areas of the polygons on the output layers which, in this example, could be used to calculate the proportion of each county covered by water.

Overlay is an essential tool that allows data to be integrated and used in new ways. It is fundamental to many of the manipulation and analysis techniques described later in this book (see Chapters 6 and 8 in particular; see also Unwin, 1996). It has also been used extensively in historical GIS research. An example of this is provided by Hillier (2002 and 2003; see also Chapter 9) in her study of mortgage 'redlining' in the 1930s. Redlining was the process by which mortgage lending companies classed the neighbourhoods into one of four types, according to how high risk they felt the neighbourhoods where. These areas were mapped with the highest risk areas being shaded red, hence the term 'redlining'. Hillier had access to these maps for the city of Philadelphia. In addition, she had two other sources of data: a large sample of mortgages granted on specific properties that she had as a point layer, and census data available for tracts. Without GIS, these data would have been incompatible, but by using overlay it was relatively easy for her to bring together these three sources, and thus perform a study of the impact of redlining areas on mortgage applications, and the further relationships with socio-economic factors, such as ethnicity.

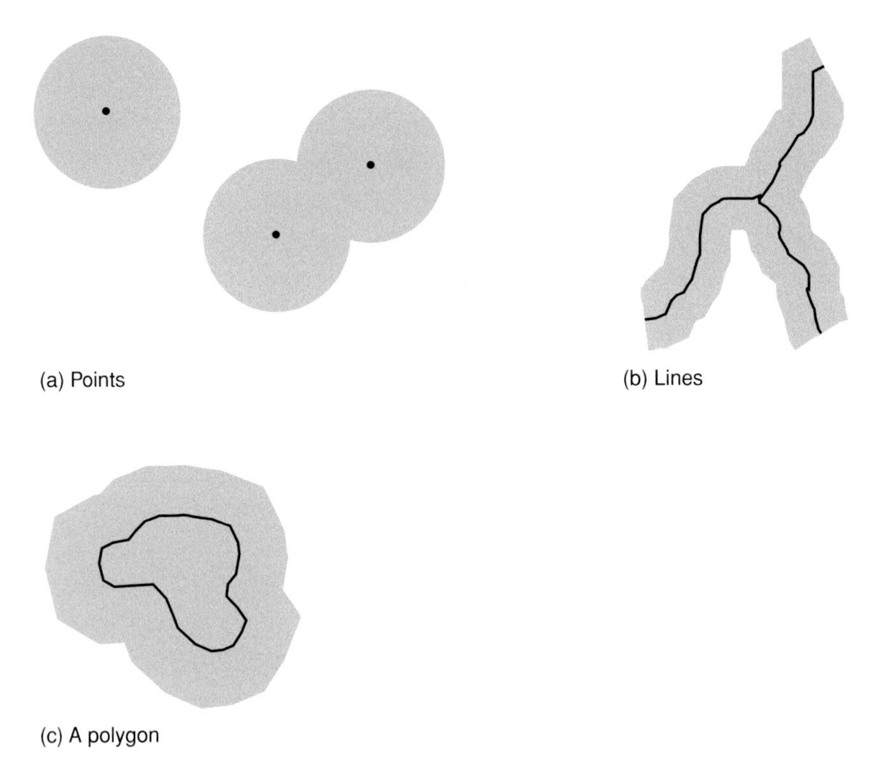

Fig. 4.8 Buffers around points, lines and polygons. The original features from the input layer are shown in black. The polygons included on the output layer are shaded gray.

Buffering is a technique that is frequently used in conjunction with overlay. As shown in Fig. 4.8, buffers may be placed around points, lines or polygons. In each case, an output layer is produced that contains polygons whose boundaries are a set distance from features on the input layer.

Usually buffering is used as a precursor to an overlay operation where it is able to assist in queries such as 'how many farms are within 5 km of a road?' Figure 4.9 shows how this would be answered. Firstly a 5 km buffer is placed around the roads layer to produce a polygon layer. This polygon layer contains an attribute that flags whether each polygon is in or out of the buffer. This new layer is then overlaid with the farms layer which adds the attributes of the buffer to each farm. In this way it is now possible to tell which farms lie within 5 km of a road. These are the ones that lie within the dotted line on the diagram.

Gregory (2007) shows an example of using buffering in historical research. He was interested in the north–south divide in infant mortality before the First World War. He argued that if there was a north–south divide, or, more properly, a core-periphery

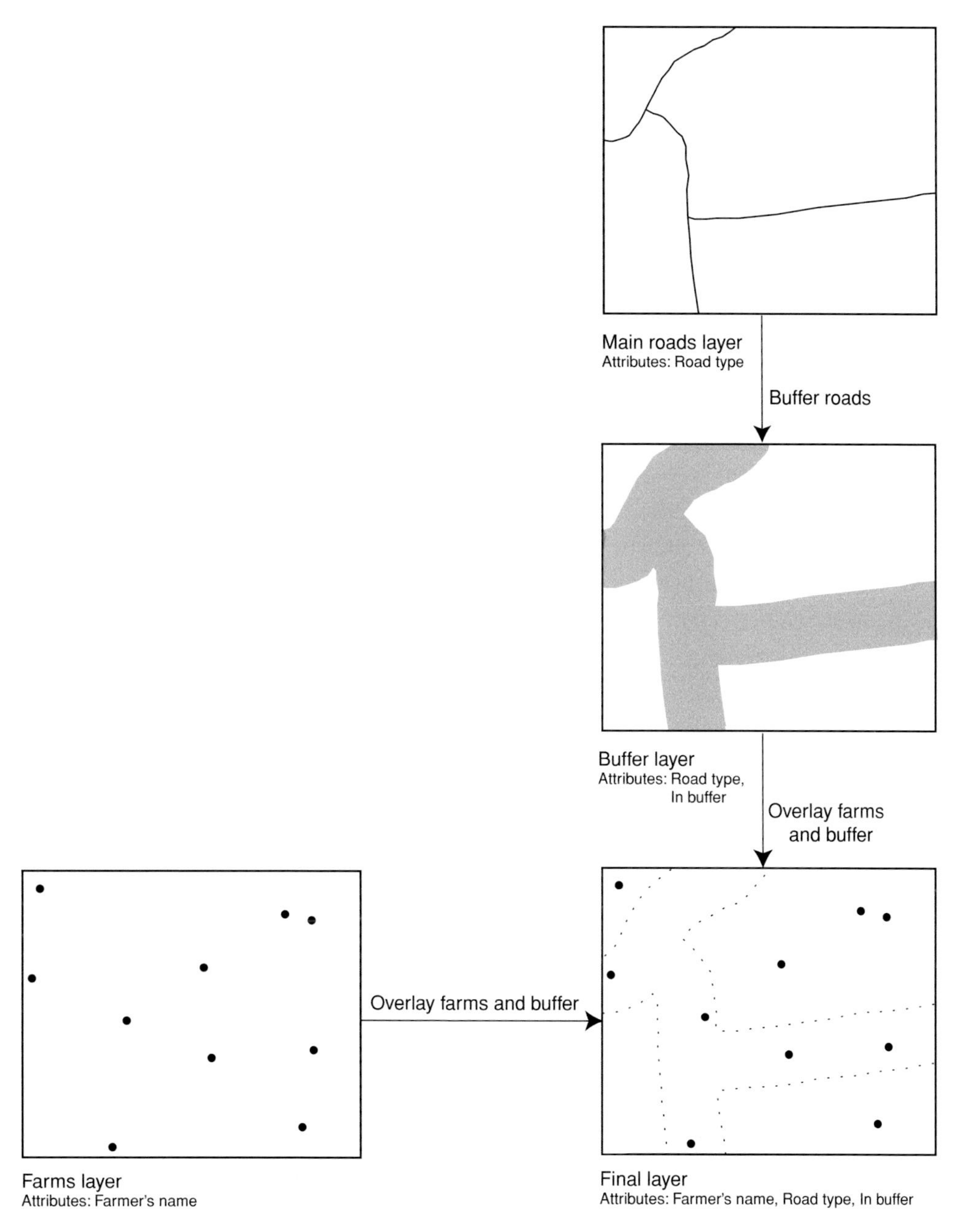

Fig. 4.9 Using buffering and overlay together. There are two input layers: a line layer representing roads and a point layer representing farms. A 5 km buffer is placed around the line layer and this is then overlaid onto the farms layer. In the output layer each farm thus has an attribute that represents whether it lies within the buffer or not.

divide, then rates should be low in and around London – the core – and increase with distance. To test this, he took infant mortality data available for polygons from the *Registrar General's Decennial Supplement*, covering the period 1901–11. This contained the number of births and infant deaths in each registration district from which the infant mortality rate (per 1,000 births) can be calculated. These were turned into a polygon layer using the Great Britain Historical GIS (Gregory *et al.*, 2002). This, in turn, was converted into a point layer where the mortality data were represented by the registration districts' centroids, rather than polygons. This is shown in Fig. 4.10a. To explore how infant mortality rates decline with distance, he took a layer containing a single point representing central London. He placed a series of buffers around this point, such that each buffer was 25 km wide. The outline of England and Wales was overlaid onto this to clip the layer, such that only the buffers that actually lay within England and Wales were left. The resulting layer is shown in Fig. 4.10b. These buffers were then overlaid onto the point layer of infant mortality data, so that the identifier of the buffer was added to the attributes of the mortality data. The points, along with the underlying buffers, are shown in Fig. 4.10c. This new layer contains enough information to allow Gregory to calculate the infant mortality rate for each buffer. Doing this involved summing the number of infant deaths and births in each buffer using the buffer identifier, and dividing this by 1,000 to give numbers of deaths per 1,000 births. These new data were then joined to the buffer layer and mapped, as shown in Fig. 4.10d.

The map clearly shows that the lowest rates are found near, but not in, London. The highest rates are found in the buffer that includes Liverpool, Manchester, West Yorkshire and some of the South Wales coalfield. This suggests good evidence for a core-periphery divide, but also shows that the pattern was more complex than this. The buffer that includes London does not have low rates, and rates in the far periphery are noticeably better than those around Liverpool and Manchester. In this way, it becomes possible to visualise how infant mortality rates vary with distance from London, and also to generate data for further analysis.

Overlay operations can also be performed on raster data, as long as the pixels on layers to be overlaid represent areas of the same size. As raster data are quite different to vector data, the nature of the overlay is also quite different. With vector data, two input layers are taken, and their spatial and attribute data are merged using a geometric intersection. With raster data, the input layers are two grids with the same pixel sizes, so the overlay simply involves calculating new values for the pixels. The handling of attribute data in vector overlay is somewhat crude – for example, in Fig. 4.6, population data from the polygons is duplicated where two or more points lie within a single polygon, so Rampton's population appears twice on the output layer,

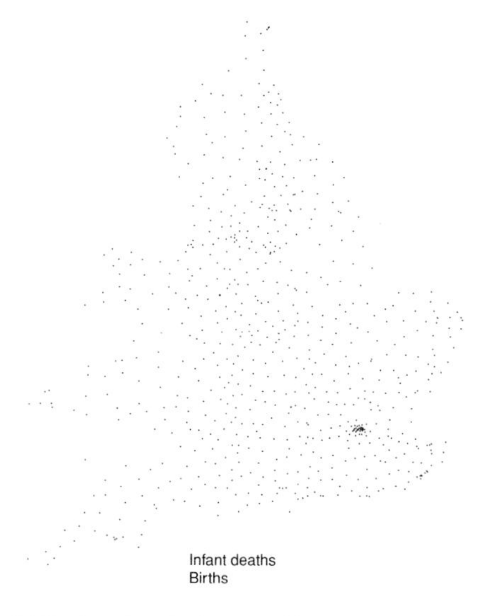

(a) Registration district centroids

(b) 25 km buffers around London

(c) Overlaying (a) and (b) produces a new point layer

(d) Summed attributes are joined back to the buffer layer and mapped

Fig. 4.10 Using buffering to explore the core-periphery divide in infant mortality in England and Wales in the 1900s. (a) The registration district centroids with infant deaths and births as attributes. (b) London with a series of 25 km buffers around it. These have been clipped using the outline of England and Wales. (c) Overlaying the centroids with the buffer gives a point layer with the combined attributes. The buffers are only shown for diagrammatic reasons. (d) Summing the infant death and birth data using 'Buff ID' and joining back onto the buffer layer allows the new data to be mapped. Legend to fig. 4.10(d): the lightest buffers hold the 10% of the population with the lowest rates: the next lightest buffers hold the second lowest decile; the highest two deciles have the darkest shading. The 60% of the population with intermediate rates is shaded in the middle shading.

Input 1

1	1	1	1	1
1	1	1	1	1
2	3	3	1	1
2	2	2	1	1
2	2	2	1	1

Input 2

0	0	0	2	2
0	1	3	2	2
0	1	3	2	2
1	1	2	2	2
1	1	1	2	2

Output

1	1	1	3	3
1	2	4	3	3
2	4	6	3	3
3	3	4	3	3
3	3	3	3	3

Fig. 4.11 Simple map algebra with raster data. The values on the output grid are the sum of the values on the input grids.

while several other parishes do not appear at all. With raster overlay, the handling of the attribute data is more sophisticated, as there are not the same issues with the compatibility of spatial areas. For this reason, raster overlay is sometimes referred to as *map algebra* or *cartographic modelling* (Tomlin, 1991).

The challenge for raster overlay is thus to take two input grids and use the values of their cells to calculate the values of the cells on the output layer. The easiest way to do this is to use arithmetic operations, such as addition, subtraction, multiplication or division. An example of this is shown in Fig. 4.11, where the values on the two input grids are added to produce the values on the output grid. Other types of mathematical and logical operations can also be performed. These include logical operations such as *is greater than* and *is less than*. These can be used to simplify a single raster surface – for example, by setting the values of all cells with a height of greater than 100 m to a value of '1', and the remaining cells to a value of '0'. In this way, cells that match the criteria 'greater than 100 m' are set to be true ('1') and the others are set to be false ('0'). This can also be used in overlay on two layers so that, for example, cells whose value on the first input grid are greater than those on the second input grid would be set to be true. The Boolean operators of *and*, *or* and *not* can also be used as part of map algebra – for example, if we have two input grids where the cell values are true or false and we want to combine these. If an *and* operation is used, only cells that are true on *both* input grids will be recorded as true on the output grid. If an *or* operation is used, cells with a value of true on *either* of the two input grids will be recorded as true. With a *not* operation, values with true on one input grid and false on the other will be recorded as true (DeMers, 2002).

Figure 4.12 shows the potential of map algebra. A user wants to identify all cells that contain either a road or a river and are less than 100 m above sea level. The first stage is to overlay the rivers layer with the roads layer using an *or* operation, such that all cells that are set to true on either of the two input layers are set to true on the output layer. This provides a layer showing all cells that contain a road or a river. Next, the altitude layer is taken and those cells with values of less than 100 m are set to true, and those above are set to false. This provides the cells whose height is less

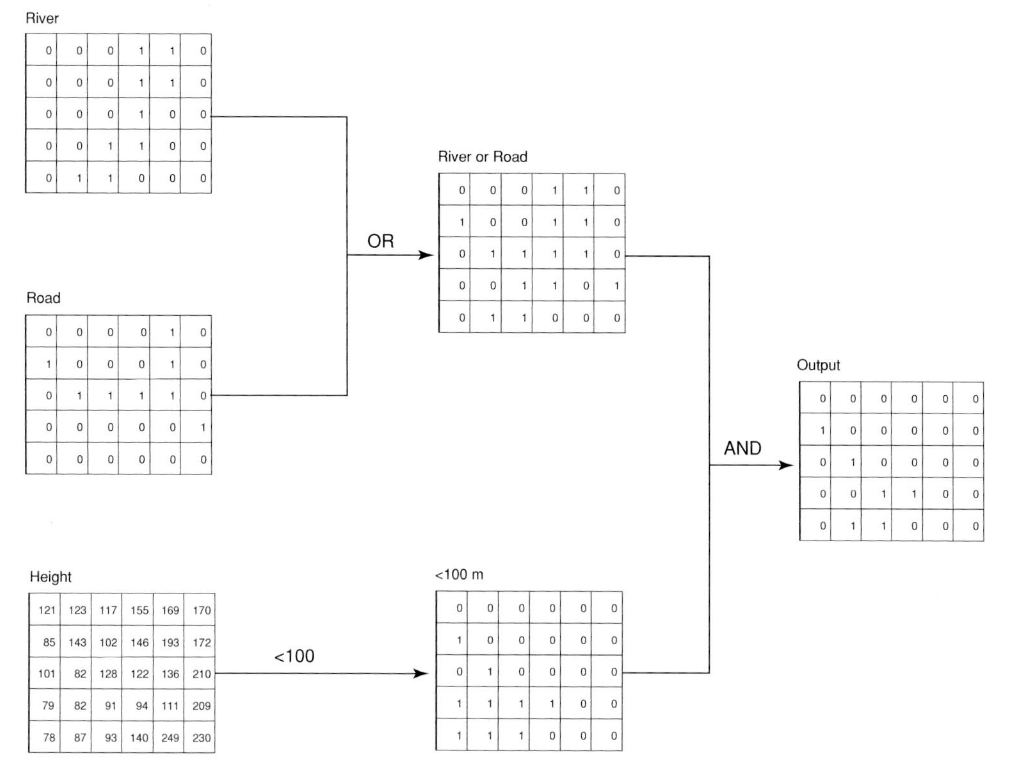

Fig. 4.12 Map algebra for site selection. The aim is to select pixels that are less than 100 m high and contain either a river or a road. To do this, three input layers are required. The *rivers* layer has cells that contain a '1' if the pixel contains a river and a '0' if it does not. The *roads* layer is similar to the rivers layer but for roads. The *height* layer records the average height above sea-level. First, the rivers and roads layer are combined using an *or* operation so that cells on the output layer contain a '1' if the cell contains either a river or a road. The height layer is then selected to produce a new layer where pixels below 100 m are recorded with a '1' and those above are recorded with a '0'. Finally, these two derived layers are combined using an *and* operation to give a layer where cells that contain a road or a river and are over 100 m are recorded with a '1'. These are the pixels that match the site selection criteria.

than 100 m. Finally, the two derived layers are overlaid using an *and* operation, such that only cells that contain a river or a road and are less than 100 m above sea level are set to true.

Map algebra, therefore, has considerable potential for deriving new knowledge from raster data. It has been used extensively in planning-type applications of modern data (see, for example, Carver and Openshaw, 1992). To date, it has not been much used in historical research, perhaps reflecting the relative lack of use of raster data; however, where data are available, there is clearly potential for this type of work.

4.5 DATA QUALITY IN HISTORICAL GIS

The tools described above allow data to be explored, manipulated and integrated in novel and powerful ways, but they make demands on the quality of the data that were probably never envisaged when the sources were created. Historians are used to considering issues of accuracy, ambiguity and incompleteness in their data, but GIS, and in particular the spatial component of data, adds another dimension to this complexity. Issues of accuracy in spatial data become even more complicated when data are integrated through overlay, as errors are likely to spread, or *propagate*, through the resulting data. There are two separate types of approaches to handling error that can be identified: technological or methodological, which looks to explicitly handle the error usually using solutions based on computer science or statistics, and interpretative solutions, where it is accepted that the data contain error and any conclusions drawn using the data are interpreted within the context of the error.

In discussing issues of data quality, it is helpful to begin by defining some terms. Unwin (1995) defines *error* as the difference between reality and the representation of it stored in the database. Only a very limited part of error should come from what he terms *blunders*, which are simply mistakes either in the source or the digital representation of it. *Accuracy* is defined as the closeness of values within the database, or results calculated using them, to values accepted as true. *Precision*, on the other hand, is the number of decimal places given in a measurement, which is usually far more than the accuracy of the data can support. The difference between accuracy and precision is particularly marked where spatial data are concerned. A point representing a city will usually have its co-ordinates expressed to sub-millimetre precision, but the accuracy of the point is probably only meaningful to within a kilometre or two. Unwin's final two terms are *quality*, which he defines as the fitness for purpose of the data, and *uncertainty*, which is the degree of doubt or distrust that the data should be regarded with. Even if a researcher starts with accurate, highly precise and good-quality data, after a number of complex operations have been performed, the resulting data may contain a high degree of uncertainty.

A major source of error is a result of the scale of the source data. Scale fundamentally limits accuracy. As was stated in Chapter 3, if a point digitised from a 1:10,000 map is 1 mm away from where it should be, either because the symbol is marked in the wrong place on the map or because of digitising error, the representation of the point in the GIS will be 10 m away from where it is on the Earth's surface. The same 1 mm error on a 1:100,000 map will be 100 m in reality. Lines give additional complications. Take, for example, a complex line, such as the coastline of western Scotland. Large-scale maps

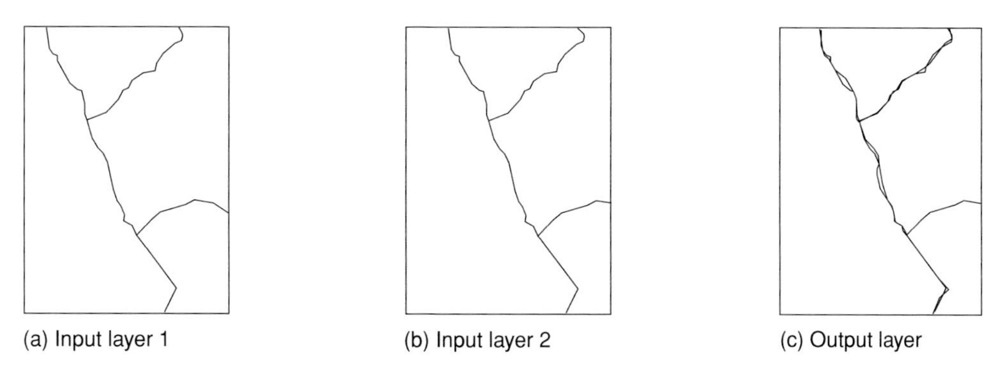

Fig. 4.13 Sliver polygons as a result of vector overlay. The diagram shows two apparently identical polygon fragments from different sources that are overlaid. As a result of minor differences in the input layers, small polygons are formed on the output layer. These are termed *sliver* polygons.

such as 1:25,000 or 1:50,000 will show many sea lochs, inlets, headlands, outcrops and other curves. As maps become increasingly small scale, many of these curves will have to be simplified or even removed entirely to allow the line to be shown on the map. This is called *generalisation*. It has the effect of shortening the line as curves are removed. On a GIS layer, it is possible to zoom in to apparently very large scales. This is misleading, as the accuracy of features on the layer cannot be higher than those on the original source. Indeed, the GIS layer will almost inevitably be worse than the original source, due to the error introduced by the digitising process (see Chapter 3).

Error becomes particularly significant in GIS when overlay operations are used. There are two aspects to this: sliver polygons being formed by the overlay, and the related issue of error propagation as increasing numbers of overlays are performed. *Sliver* polygons are created in overlay operations when minor differences in what is supposed to be the same line taken from two different sources lead to small polygons being formed in the overlay operation. This is shown in Fig. 4.13, where two apparently identical polygon boundaries are overlaid. The resulting layer contains a number of small polygons, formed as a result of the minor differences between the two boundaries. The differences that cause sliver polygons are inevitable. Even if the same operator digitises the same line from the same map to the same standards of accuracy, there will still be minor differences between the two versions. Given that overlay may involve layers that have been digitised from different source maps based on different surveys to different standards of accuracy, and perhaps at different scales, and that the digitising will also have been done to differing standards of accuracy by different people, it becomes clear that sliver polygons will be a significant problem.

There are two strategies for handling sliver polygons. The first involves removing them automatically as part of the overlay operation. Most software will allow users to do this. There are a variety of different approaches that can be used, the most common being that the user will specify a *tolerance*, or minimum distance, between any two points on the output layer. If two points are closer than this, one of them will be removed. The problem with using tolerances in this way is that if the tolerance is set too high, it will lead to polygons being removed that should not be removed. Even if this does not occur, this strategy will lead to lines being generalised as points are removed. If further overlays are required, this may lead to more problems with sliver polygons. The second strategy involves identifying and removing sliver polygons after the overlay. The easiest way to identify slivers is that they tend to be long, thin polygons, and thus have long perimeters and small areas; by dividing perimeter by area for all polygons, and selecting those with a high-ratio, potential slivers can be identified. Although this strategy is not foolproof, with some experimentation it can prove highly effective.

Once slivers have been identified, there still remains the issue of which of the two lines that forms their boundary should be deleted to remove the polygon. Some software will do this automatically, based on arbitrary criteria, such as deleting the longest line segment. It may be, however, that a user has more faith in lines from one of the input layers than in the other. Identifying this line can be done using attributes set before the overlay, but deleting the appropriate line may still prove labour intensive.

Error propagation is linked to sliver polygons, but is even more insidious. When two layers of data are overlaid, the amount of error that the output layer will contain is likely to be the cumulative total of the error from both input layers. Error in both the spatial and attribute components of the data will be affected. Error will thus spread rapidly, as a result of overlay operations (Chrisman, 1990; Heuvelink *et al.*, 1989; Heuvelink, 1999). Error propagation is more complex than sliver polygons, and there are no simple solutions to it. Instead, strategies to handle error propagation will be dealt with below, as part of a general discussion of handling error.

Error has long been identified as a key research topic in GIS (see, for example, the essays in Goodchild and Gopal, 1989, and the review by Fisher, 1999), and there have been calls for the development of 'error sensitive GIS' (Unwin, 1995). To date, however, little of this work has found its way into mainstream GIS software. One commonly proposed approach to handling error uses *fuzzy logic* (Zadeh, 1965), rather than Boolean logic, which is generally used in GIS. In Boolean logic, something is either true or false; therefore, as a result of an overlay, a point may be found to lie either inside a certain polygon or outside it. There is no room for uncertainty in this

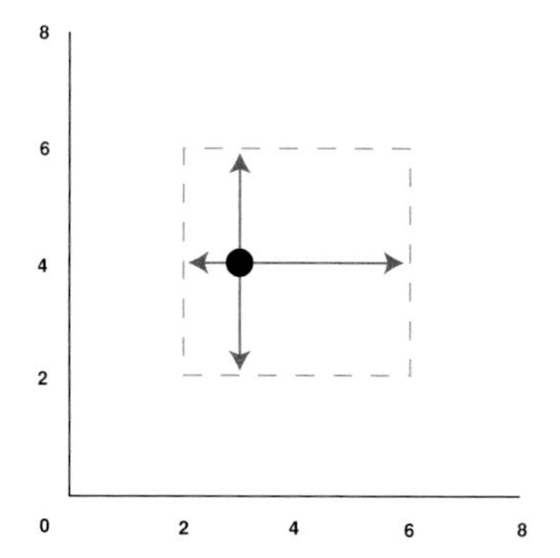

Fig. 4.14 A fuzzy point. This point is (3(−1,3),4(−2,2)) with the point of maximum expectation being at (3,4). The actual point must lie within the rectangle shown using the dashed line. Based on Brimicombe, 1998.

model. *Fuzzy logic* attempts to allow uncertainty by quantifying the probability that the point lies inside the polygon.

A relatively simple implementation of fuzzy logic within vector GIS was advanced by Brimicombe (1998). He proposes the idea of a fuzzy number, one for which, in addition to a value that is thought to be most likely, for which there is the *maximum expectation*, there is also a range of values between a minimum and a maximum. Thus a fuzzy number might be $8^{(-2,4)}$, which means that it is most likely to be 8 but definitely lies between 6 and 12. The probability of 6 is 0.0 and this rises in a straight line to be 1.0 at 8, and then declines in a straight line to become 0.0 at 12. Addition involves adding all 3 values associated with them thus:

$$8^{(-2,4)} + 6^{(-1,3)} = 14^{(-3,7)} \tag{4.1}$$

A pair of fuzzy numbers can be used to represent a co-ordinate which has a point of maximum expectation and a degree of uncertainty in positive and negative directions in both x and y. This means that a fuzzy point is represented by a rectangle on the ground. Figure 4.14 gives an example of this. The fuzzy co-ordinate is $(3^{(-1,3)}, 4^{(-2,2)})$ which gives a point of maximum expectation at (3,4) and a range of possible values between (2,2) and (6,6). Lines can be created from strings of fuzzy co-ordinates. These have a line of maximum expectation defined by the points of expectation and bands within which the line must run. This is shown in Fig. 4.15. The probabilities of the line lying at any point within these ranges, and thus which side of the line a point is likely to lie, can be calculated.

A variety of other approaches to quantifying uncertainty have been proposed (see, for example, Goodchild *et al.*, 1994; Goovaerts, 2002; Kollias and Voliotis, 1991);

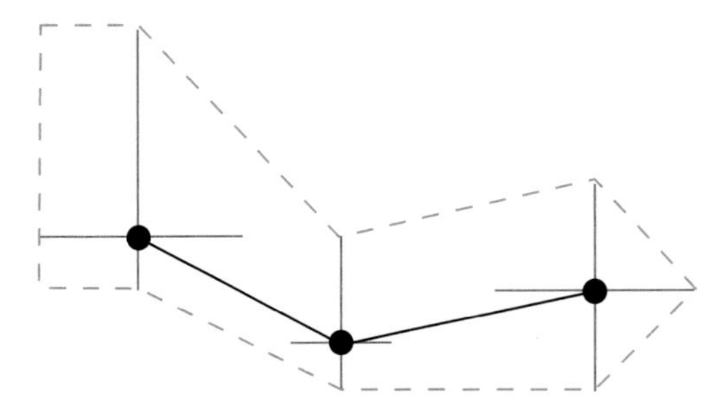

Fig. 4.15 A line defined by fuzzy co-ordinates. The bold line is the line of maximum expectation, while the line must lie within the dotted area defined by the fuzzy co-ordinates. Based on Brimicombe, 1998.

however, these have not been adopted by GIS software, and their relevance to historians may be questioned. There is one good example of how fuzzy set theory can be applied to uncertainty in historical Geographical Information. This is developed by Plewe (2002) as a method of developing a database of the evolution of county boundaries in the state of Utah, where, in the early years, sources are limited and frequently ambiguous. He uses fuzzy set theory to develop an Uncertain Temporal Entity Model (UTEM) that describes the causes, types and nature of the uncertainty contained in his database. This is able to handle spatial, temporal and attribute uncertainty. Spatial uncertainty arises, for example, when boundary descriptions state that the boundary runs along a ridge line, but this is hard to define precisely on the ground. An example of temporal uncertainty is where there is doubt about when a boundary was changed. Attribute uncertainty occurs, for example, where there is doubt about the population of a county. In each case, the model explicitly records what the uncertainty is and what its impact may be, so that it is possible for the user to understand the limitations of the representations used in the GIS. It, therefore, effectively combines the conventional spatial and attribute data with metadata on the uncertainty that these contain. The model that Plewe develops is conceptually thorough, in that it handles all forms of uncertainty; however, the practicality of developing models of this sort, and their utility in further research, remain to be proven.

Although the GIS literature focuses on statistical solutions to error and uncertainty, historians are well used to handling these problems using an entirely different approach, namely through documentation and interpretation. Chapter 3 deals with issues of documenting datasets. Given adequate documentation, it becomes possible for the historian to consider how error and uncertainty are affecting a database and, in particular, influencing the results of any analysis that the historian may want to perform. Superficially at least, this is more labour intensive than the more automated approaches described above; however, it means that the historian is focused directly on how error and uncertainty affects the analysis, rather than lots of effort being put into sources of error that may not be relevant. This does, of course, rely on suitable metadata being present, and on historians interpreting the geographical as well as the historical elements of the uncertainty present. This is by no means a solution that is able to handle all eventualities – for example, where multiple overlays have been performed – but is perhaps a solution that is more in line with traditional historical approaches than the ones described above.

4.6 CONCLUSIONS

The GIS data model and the way that it combines spatial and attribute data provides an unparalleled way of exploring, structuring and integrating spatially referenced data. These stress the spatial nature of data in ways that no other representation – be it a database or on paper – can manage. This, in turn, provides new methods and new tools to re-examine questions in historical geography that have traditionally been hampered by the complexity of their data. Within GIS, the use of the map allows the user to structure, query, integrate and manipulate data in a way that is both explicitly geographical and highly intuitive.

The potential for this can be illustrated by MacDonald and Black (2000). Although they are interested in the history of print culture, or book history, their arguments are relevant across historical GIS. They present the abilities of GIS, as shown in Fig. 4.16. The point that is implied by this diagram is that, within historical GIS, the aim is to attempt to re-create a study area that no longer exists. The only way that we can attempt to do this is by using sources that still do exist. These will inevitably be fragmentary and disparate, collected by different organisations in different ways and for different purposes. In their example, they point to the importance of transport routes as a conduit to allow the influence of books and literacy to spread. Information on these is likely to be available from contemporary maps. Demographic statistics will be available from the census, published using administrative units. Information that is more specific to book history, including book production, book stores and libraries, is available from a variety of sources, including trade directories and official reports. Again, these may have spatial references, such as the names of towns or, perhaps, specific addresses whose location can be determined. The only way that this

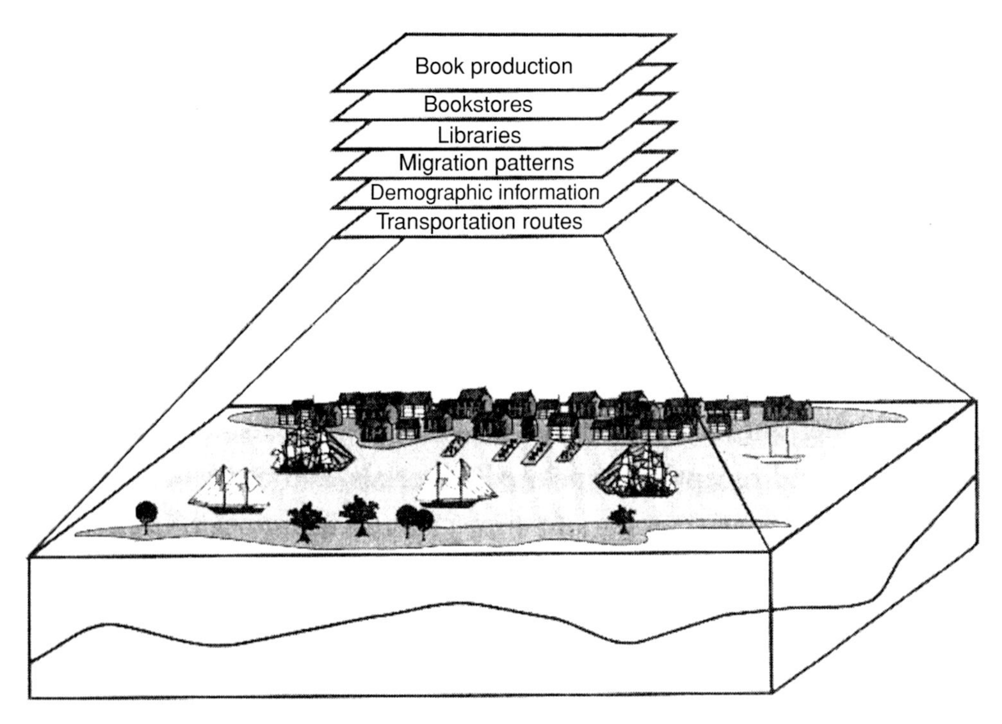

Fig. 4.16 Using GIS to re-united disparate data sources. An example from the history of print culture. Source: MacDonald and Black, 2000.

information can be brought back together is through location, and GIS provides a highly effective tool for doing this. As such, it has the potential to be of much benefit to historical geographers.

Having said this, there remain many issues to be resolved. Primary among these is that, while GIS allows data to be brought together at a technical level, interpreting it in a scholarly way is still problematic. To follow the example of book history, we may know where libraries are located as points, and know about the distribution of people and their literacy levels from the census. Overlay allows us to find out which census areas the libraries lay within, but interpreting what a library's catchment area is, and which people from within the catchment were likely to use the library, remains problematic. This is not to say that it is impossible to do this; merely that GIS provides technical solutions, but it is still up to the researcher to solve scholarly ones. Much the same situation exists with the problems of error and uncertainty. While there are some possible technical solutions, these remain, in many ways, a scholarly issue that can only be fully resolved with solutions involving careful study and interpretation.

5

Using GIS to visualise historical data

5.1 INTRODUCTION

While many historians regard GIS as being substantially about mapping, it is hoped that, having reached this stage of this book, it is clear to the reader that GIS is about far more than this: GIS is effectively a spatial database technology concerned with structuring, integrating, visualising and analysing spatially referenced data. The map, and visualisation more generally, is the most readily accessible way of communicating the data held within the GIS and the results from any GIS analysis. It is, however, a huge mistake to equate GIS and mapping as one and the same. If a researcher is simply interested in creating a number of maps as part of a research article, then conventional cartography – probably using a computer – is likely to be the best way to proceed (Knowles, 2000). There are two reasons for this: firstly, the length of time that it takes to create a GIS database, and secondly, GIS tends to lead to rather formulaic maps based around points, lines and polygons, whereas traditional cartography, whether or not it is done on a computer, allows for more freedom of expression.

Nevertheless, GIS and mapping are inseparable. This chapter has three main aims: firstly, to describe how to produce and interpret the types of maps created by GIS, secondly, to evaluate the extent to which mapping advances our understanding of historical research questions, and thirdly, to go beyond simple mapping, to more complex electronic visualisations, such as cartograms, animations, virtual worlds and the use of electronic publication. The main message of the chapter is that maps and other visualisations need to be treated with caution. They have the ability to present deceptively simple patterns, either deliberately or through poor cartography, and they are often attractive products that appear to be meaningful, but tell us very little. They are, fundamentally, illustrations of the data held within the GIS, which means that they are limited as analytic tools. With careful handling, they are useful in helping to understand the spatial patterns within a dataset. However, they only provide a limited view of the geography within the data, and their importance should not be overstated. Fundamentally, they are good at illustrating a story, but poor at telling it.

5.2 GIS AND MAPPING: STRENGTHS AND WEAKNESSES

GIS combines spatial and attribute information, and the map provides an effective way of presenting this in an understandable manner. Mapping opens the possibility of exploring the spatial patterns inherent within the data in ways that were never previously possible. The potential of this can be demonstrated by looking at Spence's (2000b) work on late seventeenth-century London. He has a single layer of data that contains the administrative geography of London in the 1690s. He links this to a selection of taxation data, and uses the resulting maps to explore the various aspects of the geography of London at this date, including business rents, household density and rents, the distribution of households by gender, and so on. Finding spatial patterns in these data allows him to advance possible explanations for them – for example, he finds that there were high concentrations of widows in the City and in the East End. He suggests that there are different explanations for each of these. In the City, he argues that it reflects the dominance of men as property owners. There was a very low proportion of female property holders in the City, and in the majority of cases where women did own property it was because they had inherited it from a dead husband. In the East End, however, he argues that it reflects the poverty in which many people lived and that males, in particular, were vulnerable to premature death as a result of the dangerous types of employment available to them.

This illustrates both the strengths and weaknesses of a map-based approach within GIS. Spence has several types of attribute data which he is able to link to a single layer of spatial data, namely the administrative boundaries. Once he has done this, he is able to describe the patterns that the maps show, and this stresses the spatial variations found across the city. It does not, however, explain the patterns. This requires more conventional historical scholarship. The use of maps, therefore, presents a challenge to the historian, as it demonstrates the patterns within the data and challenges him or her to explain them. It does not provide the explanation, and there is a danger of jumping to easy conclusions that may not be justified by more detailed research. As Baker (2003) says, 'mapping is fundamentally descriptive and proactive, rather than necessarily being interpretative and productive. While it might answer the question "Where?" it does not in itself answer the question "Why there?" Indeed, far from answering this latter question, mapping distributions itself raises that very question' (p. 44).

In recent years, the nature of maps and cartography has attracted a certain amount of criticism (Harley, 1989 and 1990; Pickles, 2004). Maps are often regarded, both by map readers and cartographers, as objective and scientific representations of reality that provide 'a mirror of nature' (Rorty, 1979 quoted in Harley, 1989: p. 4). The choice of what is displayed on maps and how it is displayed, however, is highly subjective. It is controlled by the values of the cartographers that produced the maps, and the culture that they originate from. The map maker selects, omits, classifies and symbolises in ways that tell the story that he or she wishes to tell. Thus,

> All maps strive to tell their story in the context of an audience. All maps state an argument about the world and they are propositional in nature. All maps employ the common devices of rhetoric such as invocations of authority . . . and appeals to the potential readership through the use of colours, decoration, typography, dedications, or written justifications of their method. Rhetoric may be concealed but it is always present (Harley, 1989: p. 11).

Regardless of these criticisms, maps produced by cartographers can be regarded as a fusion of scientific surveying and measuring combined with the art of presentation, to summarise complex information about the Earth's surface. They are highly complex pieces of work created by skilled craftsmen and women using centuries of accumulated knowledge. At their best, they are works of art that rival paintings; indeed, there is a thriving market in antique maps that is in many ways similar to the art trade. The maps produced by GIS are a far cry from this. They are usually thematic maps that use symbols or shading. How this is done will be returned to in detail in the next section, but, for now, let us simply consider how well these maps represent the three components of data: space, attribute and time. In introducing GIS in Chapter 1, we argued that, traditionally, to explore one of these three components, we need to simplify the second and fix the third. This, we argue, is unsatisfactory, and in GIS we should be attempting to make full use of all three components simultaneously. How well does a thematic map allow us to do this?

Figure 5.1 shows a typical example of the kind of *choropleth map* easily produced by GIS. In this case, the map shows deaths from lung disease amongst males in the second half of their working lives, in the 1860s. The attribute data are taken from the *Registrar General's Decennial Supplement*, and they have been linked to spatial data showing the boundaries of Registration Districts taken from the Great Britain Historical GIS (Gregory *et al.*, 2002). The map shows that the highest rates tend to be found in the major urban and industrial centres, such as Liverpool, Manchester,

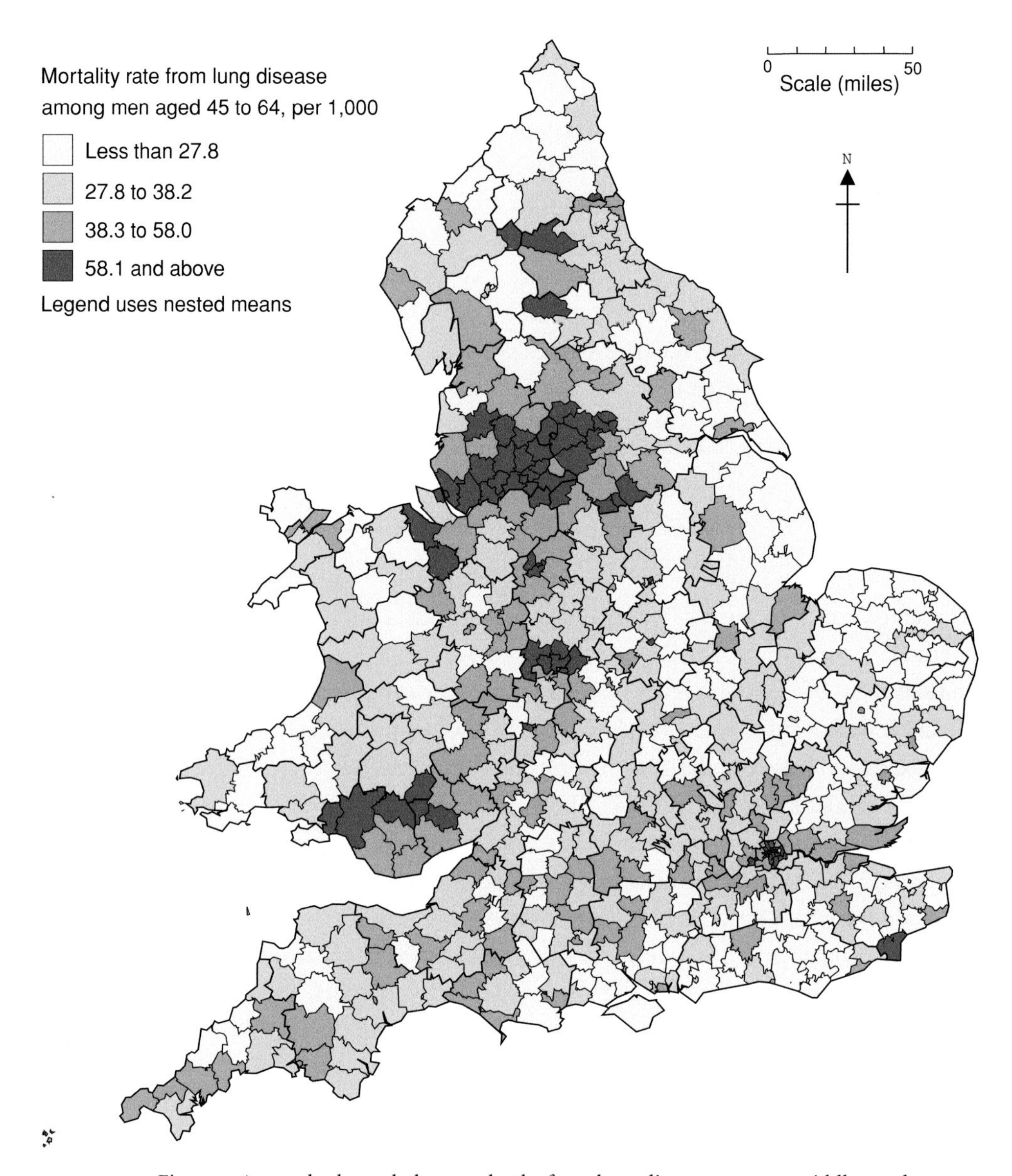

Fig. 5.1 A sample choropleth map: deaths from lung disease amongst middle-aged men in England and Wales from 1861 to 1871. The map shows the mortality rate per 1,000 males aged 45–64 – the second half of their working life – from lung disease. Source: *Registrar General's Decennial Supplement*, 1871.

the West Yorkshire area, Birmingham, and the South Wales coalfield. Low rates are concentrated in the east of England. There are a variety of outliers to the pattern – for example, the three large rural areas in the north of England, and Romney Marsh in Kent. It might be speculated that some of these are due to coal mining, although the map does not, and cannot, confirm this.

Thematic maps such as this stress the spatial, and allow us to make statements like the ones above that describe patterns, and speculate on causes. To enable them to stress the spatial pattern, however, attributes have to be simplified. In choropleth maps where polygons are shaded according to their attribute values, this is usually limited to around four or five attribute classes (see section 5.3); in Fig. 5.1 only four have been used. With point symbols, more detail is available, but the number that can be distinguished by the reader is still limited (Robinson *et al.*, 1995). These classes have to be defined in an arbitrary way. Although there are guidelines about how to do this, described in section 5.3, the choice is fundamentally subjective, and different choices may result in very different patterns being mapped from the same dataset. Therefore, to stress the spatial, attribute needs to be highly simplified. Time in a thematic map is effectively fixed. It is possible to compare two different dates on one map or to turn time into an attribute and simply show where something occurred at different dates using dot maps. When publishing electronically, animations can be used to show change over time; however, as will be discussed in section 5.6, this requires further simplification of attribute and space to make them understandable. In general, therefore, time is fixed on a paper thematic map.

Even choropleth maps of administrative areas can be problematic. Administrative units tend to be large in sparsely populated rural areas and small in dense urban areas. This means that the map tends to be dominated by rural areas in which the fewest people live while very densely populated areas, such as big cities, can all but disappear. In Fig. 5.1, the three large outliers in the north of England – Alston, Weardale and Reeth – seem significant, but only have a combined population of around 30,000. Two much smaller outliers, Worcester to the south of Birmingham, and Leicester to its east, have populations of 32,000 and 95,000 respectively, but appear far less important. It is very difficult to make any statements at all about the pattern in London, where around 15 per cent of the population live, as its small size makes it almost indistinguishable. Again, solutions to this, such as cartograms, can be developed, and these will be discussed below at section 5.5, but it is fair to say that space in a thematic map is often deceptive, and may be misleading. A final issue with choropleth maps in particular, is that they can generally only show one theme at a time, and it is thus very difficult to see how the pattern of two or more variables inter-relate.

The maps produced by GIS are thus very limited products. In some ways, it is constructive not to think of them as maps at all, but instead to regard them as a form of graph not unlike a histogram or a scatterplot. They provide a simple summary of the distribution of a variable in ways that attempt to stress the spatial by simplifying attribute and fixing time. The ways that space is represented and that attribute is simplified can both lead to distortions in the way that the map reader perceives the underlying pattern. This occurs even when good cartography is used, and, frequently, as will be discussed below, GIS users are not good cartographers. A map in GIS is therefore no more than a simple and distorted summary of an underlying dataset which is in itself a simple and error-prone abstraction of the real world. As such, it must be interpreted with caution.

5.3 CREATING GOOD MAPS: CARTOGRAPHY AND GIS

Having said, in the previous section, that a GIS map may be regarded as a form of diagram, there is at least one very important difference: diagrams attempt to provide an objective summary of a dataset. Maps, even the simplest maps from a GIS, are more complex than this, and rely on a large number of subjective and arbitrary choices by the map maker to present the message that they tell. These choices will significantly impact on the story that the map tells to the reader. This leaves a heavy responsibility with the map maker to make informed choices in the way that the map is designed, and also with the map reader to ensure that they critically evaluate the message that the maps presents to them, rather than regarding it as some form of objective representation of reality. An excellent discussion of how maps can present deceptive messages, both through deliberate cartographic choices and accidentally through poor cartography, is provided by Monmonier (1996).

Many good guides to cartography are available (see, for example, Dorling and Fairburn, 1997; Keates, 1989; MacEachren, 1995; Monmonier and Schnell, 1988; Robinson *et al.*, 1995), and it is not the intention here to provide a full guide to cartography. Instead, a brief introduction to map design will be given to provide some basic information on how to both produce and critique the types of maps frequently produced by GIS.

The basic challenge to the GIS user wanting to produce an effective map from their data is to take a complex dataset and simplify it enough to make the patterns within it understandable on the printed page or the computer screen, without over-simplifying the data or distorting the patterns that it contains. The map, therefore, needs to present an honest summary of the data. Too little simplification will result in the patterns shown not being understandable. Too much simplification, and the patterns will either disappear or be too simplistic to be meaningful. The first choice that this imposes on the map maker is what data to include on the map. This is usually a highly subjective choice, but may be strongly led by the data themselves. If shaded polygons are to be used, it is usually only possible to show one variable from one layer of data. Dot maps provide for more flexibility in choice, and it may be that several layers can be included on a single map. It is also possible to use pie charts on dot maps to show more than one variable from a single dataset.

Let us take the example of mapping data on the occupational structure of towns, where each town is represented by a dot whose size is proportional to the town's population size, and the proportion of its population employed in each employment type is represented by subdividing the dots into slices whose size depends on the proportion of the population who work in each employment type. If too many employment classes are used, it will be impossible to distinguish one class from another; if too few are used, the information on the map will not be distinguishable.

There are few hard and fast rules about these choices that can be applied to all datasets presented on all mediums. Decisions will depend on whether the map is to be published in grey-scale or in colour, as the use of colour allows the human eye to distinguish more symbols or classes. The number of observations in the dataset is also important, as more values will generally lead to a more complex pattern, which in turn will mean that the number of classes of employment types needs to be simplified. One key point is that the number of shadings that can be distinguished by the human eye is surprisingly limited. In grey-scale, four or perhaps five is the maximum that human perception can distinguish, and even this relies on good quality printing. Colour allows for a few more classes to be distinguishable. Ten is sometimes referred to as the absolute maximum, but this is optimistic. The choice of numbers of classes is thus both subjective and limited. The map maker needs to choose carefully, but must be aware that it is always tempting to include more classes; however, this may lead to the pattern becoming over-complex. Another issue is the choice of whether to use background information to provide context to the data. It may be desirable to use raster scans of historical maps, terrain models to show relief, or vector layers showing transport routes, rivers, settlement locations or other information that helps the readers to orientate the data on the map. Used well, these can add to the message that the map presents, and they may, in fact, be essential. Used badly, however, the background information will provide too much detail which will distract from or obscure the main message of the map.

A second major issue that a map maker should consider is that the map should be a stand-alone document. To understand it, the reader should not have to refer to the accompanying text. To allow this to happen, the map needs a title, a legend that explains all of the symbols used on the map, and a scale bar. All required information – for example, sources of data and choices of class intervals – should be presented on the map itself.

Careful choices of symbols and shading can help to achieve this aim. Even though a legend is essential, the choice of symbols and shading should be done in such a way that it attempts to allow the reader to understand the map without referring to the legend. In reality, although this is unlikely to be completely achievable, many choices can be made that assist the reader. There are many symbols that are internationally recognisable, for features as diverse as churches, lighthouses, railway lines, shipwrecks, and so on. Similarly with colours, blue often represents water, brown may imply urban, green may be rural or may also signify safety, while red tends to imply importance or danger. Symbol size or intensity of colour can also be used to imply importance. The human eye tends to be drawn to large symbols and intense colours at the expense of smaller symbols and more subdued colours. The intensity of colours is easy to control in GIS or computer cartographic software. The computer builds up a large number of colours from three or four basic ones. These are usually either red, green and blue combinations – called RGB combinations where the map maker can vary the amount of each of these three colours to produce any colour, or CMYK combinations which work in the same way, but use cyan (light blue), magenta (light purple), yellow and black. Using small amounts of all three or four colours leads to subdued shades, while using large amounts of one or all of them will lead to intense colours. This means that part of the map maker's art is to use the size of symbols, choice of colours, and intensity of shading to draw attention to the aspects of the map that he or she considers important. This may be to emphasise big cities rather than small villages, or areas with high unemployment rather than low. Part of the map reader's art is to accept that this is happening, and to make sure that they are not deceived by it.

A specific aspect of the use of intensity is to emphasise hierarchy. Many GIS maps represent *ordinal, interval* or *ratio* data. Ordinal data are classed into types with a pre-defined hierarchy. This may be found, for example, with data on social class, which in the British census are classed from I to V, with I having a professional head of household, and V having an unskilled worker as the head of the household. With interval or ratio data, the data are numeric, and need to be subdivided into classes for mapping, using techniques described later in this section. The difference between ratio and interval is that with ratio data there is a meaningful zero, whereas with interval there is not. For example, a district with a population of 200 has twice as many people as a district with a population of 100, so population is ratio data. Temperature is interval data, as 20 °C cannot be said to be twice as hot as 10 °C.

Regardless of the type of data, where a hierarchy exists it is important that this is reflected in the shading scheme, such that the reader can tell what order within the hierarchy each shading lies without looking at the legend. This is achieved by careful choice of both colour and the intensity of shading. The default shading schemes offered in many GIS software packages often do not do this well. An excellent choice of colour schemes is provided by Brewer (2002). In addition to providing many RGB or CMYK combinations, this site gives an indication of how well they work on different media, such as photocopies, computer screens and data projectors. It also provides information on how well each combination works for people who are colour-blind.

A final but very important issue is the choice of class intervals that are used to subdivide interval or ratio data, particularly on choropleth maps. As with colour schemes, the fact that GIS offers default options means that it is tempting for users to simply accept these. In reality, they should be making an informed and thoughtful decision which involves making a number of choices. The first of these is the number of classes to be used. As was discussed above, this is usually made based on the complexity of the data, the number of polygons, and the type of output media to be used. More than five classes are rarely easily distinguishable by the human eye. This means that the attribute data are heavily simplified, and this must be done sympathetically to the distribution of the data. It is important that some characteristic of the data is used to determine the class intervals, rather than simply using the ones that the map maker thinks gives the best pattern. Therefore, if a reader sees a choropleth map with classes such as 0 to 4, 5 to 24, 25 to 34 and 35 to 40, these should be treated with extreme caution, unless a good reason is given for this choice, as, in all probability, these have been selected to show the pattern that the map maker wants to show, rather than a pattern that genuinely appears in the data.

There are many other approaches to defining class intervals in a more objective manner. A seminal paper by Evans (1977) describes a wide range of these. The key points are not to use arbitrary values selected to give the best pattern, and to think carefully about which choice to use, bearing in mind the characteristics of the data. In some cases, there is a characteristic of a dataset that should be explicitly included into the selection of class intervals. Examples include a population change of zero, or a sex ratio of one male to every female. It may not always be appropriate to actually use this value as a class break, but rather to have it in the middle of a class which shows, for example, areas with very low population change or areas with approximately the same numbers of men and women.

In most other cases, the frequency distribution of the variable will give some indication of how to define class intervals. Where the frequency distribution shows a normal distribution, there is considerable flexibility. One option is to take a statistical approach and to use, for example, the mean as the central class break, and the standard deviation to define classes on either side. Where data have a slightly skewed distribution, the use of nested means may be appropriate. This involves first using the mean to divide the dataset into two classes. These classes are then both subdivided using the mean of their observations to give four classes. Another option is to use equal intervals, whereby the range of values is calculated by subtracting the maximum value from the minimum and this is divided by the number of classes that are wanted. For example, if a dataset has a maximum value of 25 and a minimum of 5 and the user wants to create four classes, the intervals become 5, 10, 15, 20 and 25. This, again, works well with data that are slightly skewed.

An alternative approach is to allocate equal numbers of observations to each class. Therefore, if there are two hundred polygons, and the map maker wants to create four classes, the intervals will be defined to put fifty polygons into each class. When this is done with four classes, the median becomes the central class break and the lower and upper quartiles are the other two. Again, this can be an effective option with slightly skewed data. One problem with this approach is that it gives equal weight to each observation, which, if the polygons represent administrative units with very different population sizes, may be misleading. An alternative is to attempt to give each class an equal proportion of the population of the study area, such that, for example, each class gets 25 per cent of the population, rather than 25 per cent of the observations. A slight variation on this is that if the data are concerned with inequality across a population, poverty researchers often split the population into ten deciles. Rather than shade all ten, which is too many classes, the two lowest deciles can be shaded using two shades of one colour, the highest two with two shades of a contrasting colour, and the middle six shaded in a pale colour, as these are seen as less important. This emphasises the difference between the best and worst areas, taking into account their different population sizes.

Heavily skewed data, such as population density across a country, are the hardest to subdivide into classes satisfactorily, as there are a large number of very small values and a small number of very large values. Using equal numbers of observations is one option, but this tends to subdivide the small numbers into many categories, while not subdividing the higher values. Instead, an arithmetic or geometric progression may be used. The aim of an arithmetic progression is to start with the minimum value, say 1 and increase this by an increment value, say 3. This gives a first class interval of 4. We then add 3 to the increment value to give $3 + 3 = 6$ and add this to

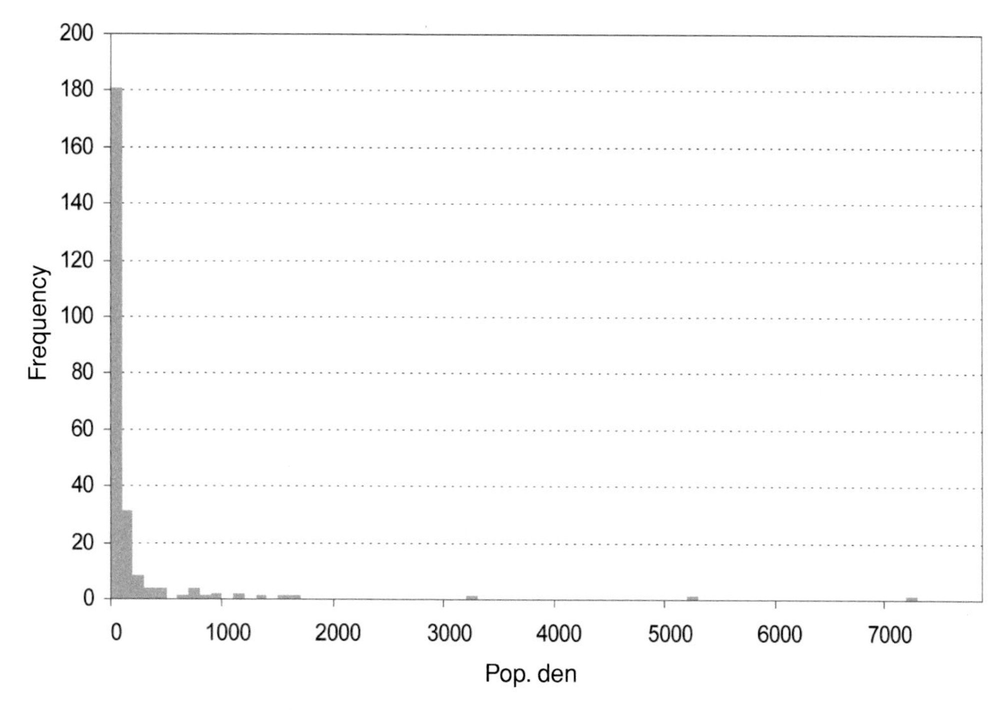

Fig. 5.2 Frequency distribution of population density in Warwickshire parishes in 1951. Population density is in persons per square kilometre. Sources: Population taken from the 1951 census reports. Areas as calculated by the Great Britain Historical GIS.

the first class interval giving a second class interval of $6 + 4 = 10$. Again, 3 is added to the increment, meaning that we increase by $6 + 3 = 9$, giving a class interval of $9 + 10 = 19$. Thus we have a progression of 1, 4, 10, 19, and so on. One problem with this is how to determine the increment value. The best way to do this is to choose a value that will start at the minimum data value and finish at, or close to, the maximum value within a reasonable number of classes. A geometric progression is similar to this, except, rather than adding the increment value every time, it is multiplied. Thus starting at 1 with an increment of 3 gives a class break at 4, as above, but then we add $3 \times 3 = 9$ to the increment to give a class break of $9 + 4 = 13$ (compared to 10 in the arithmetic progression). $9 \times 3 = 27$ which gives a class interval of $27 + 13 = 40$ (compared to 19), and so on. This gives a very rapidly increasing scale.

Figure 5.2 shows a frequency distribution of parish-level population density in the county of Warwickshire in 1951. The distribution is heavily skewed, with around half of the observations being less than 50 persons per km^2 but two observations, Birmingham and Stratford upon Avon, having values in excess of 5,000 persons per km^2. The range of values goes from a minimum of 2.8 to a maximum of 7,285.6

persons per km^2. The data are mapped in Fig. 5.3 using four different approaches. In Fig. 5.3a, nested means are used. Figure 5.3b uses an equal number of observations in each class, giving a much wider highest class than Fig. 5.3a. Figure 5.3c, by contrast, uses an arithmetic progression. This starts at zero and increases by 800, giving a top class of 4801 to 8,000, slightly above the maximum value of the dataset, but quite close. The problem with this system is that the lowest class is too wide and thus contains too many observations. Finally, in Fig. 5.3d, a geometric progression is used that starts at 0 and increases by 9. This gives a maximum value of 7,380, very close to the maximum value. This scheme, arguably, puts too few observations in the lowest class, but represents the highest values quite well. It may be argued that nested means give the best representation of the data in this case, but this is a subjective judgement and one that can only be made by comparing the different patterns. Attempts have been made to provide more objective measures of the closeness of fit of class intervals to the underlying data that they are trying to represent (for example, Jenks and Caspall, 1971), but whether these improve on the subjective judgements is open to question.

Trying several possible ways of selecting class intervals is to be encouraged. While this can lead to accusations of presenting the data in the best possible way, it encourages the map maker – which in GIS terms is the researcher – to explore the dataset thoroughly and think about why different patterns appear using different intervals. When a map is published, the choice of class intervals should always be included with the map.

In short, therefore, the map maker must make a number of subjective and artistic judgements. He or she should be aware of the impact of these on the pattern shown, and should use them to try to tell the story that they feel is present in the data, or to emphasise the points that they want to emphasise. The map reader should not regard the map as an objective work, but should instead see it as a subjective presentation of the data that attempts to demonstrate it in a certain light. He or she should therefore think critically about how design choices have affected the patterns that appear.

5.4 HISTORICAL ATLASES MADE USING GIS: A GOOD USE OF GIS?

Given the links between GIS and mapping, it is not surprising that GIS has been used extensively to produce atlases. Two distinct approaches can be identified to this. In the first instance, GIS technology has been used to produce what might be referred to as traditional atlases, using the technology to reduce the production costs. Although GIS has been used, the finished product looks indistinguishable from a conventional

Fig. 5.3 Four different views of the same data. The map shows population density in Warwickshire in 1951 classed using (a) nested means; (b) equal observations; (c) an arithmetic progression starting at 0 with an increment of 800: (d) a geometric progression starting at 0 with an increment of 9.

atlas. The second type are atlases that are produced from GISs, usually by taking a small number of layers of data, typically administrative boundaries, and linking them to a wide range of attribute data concerned with different themes. These are distinctly GIS maps, and tend to be more formulaic and less artistic than the former, being based primarily on choropleth or other thematic types of maps.

An example of using GIS to produce traditional atlases is provided by the Historical Atlas of Canada project. This was a three-volume series covering the period up to 1800 (Harris, 1987), the nineteenth century (Louis, 1993), and the twentieth century (Kerr and Holdsworth, 1990). The project was proposed in 1978 and was large in scope and generously funded. In spite of this, the sheer volume of data and maps and the large number of authors involved led to the project getting behind schedule and over budget. At this time, there was no serious consideration of using GIS, as the technology was still in its early stages. By the time Volume II, the last of the three volumes to be completed, came to be produced, GIS was considered as a serious option and was used in the production of many of the maps. The project employed professional cartographers, and continued the standards and approaches used in the earlier two volumes. The quality of the cartography was very high, and the atlas focuses on distinct historical themes. The advantage of using GIS was largely in keeping costs down (Pitternick, 1993).

Atlases produced as ways of publishing GIS data tend to look very different from the Historical Atlas of Canada and other traditional atlases. Examples include Kennedy *et al.*'s (1999) atlas of the Great Irish Famine, Woods and Shelton's (1997) atlas of mortality in Victorian England and Wales, and Spence's (2000a) atlas of London in the late seventeenth century. All of these follow the same basic formula. The authors start with a very small number of layers of administrative boundaries: Woods and Shelton have a single set of registration districts for England and Wales, Spence has a single set of parish boundaries for London in the 1690s, and Kennedy *et al.* have a number of sets of boundaries representing the major Irish administrative units between 1841–71. All the authors' atlases link their boundaries to rich sets of attribute data: Woods and Shelton use an extensive range of mortality data drawn primarily from the *Registrar General's Decennial Supplements* for the Victorian period, Spence has a large amount of data taken from taxation records for the 1690s, and Kennedy *et al.* use a large amount of census, Poor Law and related data. This means that, based on simple spatial data, the authors are able to produce a wide range of maps on different themes associated with their central topic. Figure 5.4 shows an example of this, taken from Kennedy *et al.*'s atlas. It uses three maps from 1841, 1851 and 1861 to show the impact the Great Irish Famine of the late 1840s had on poor-quality housing. The census recorded housing in four classes, with class four being the lowest which

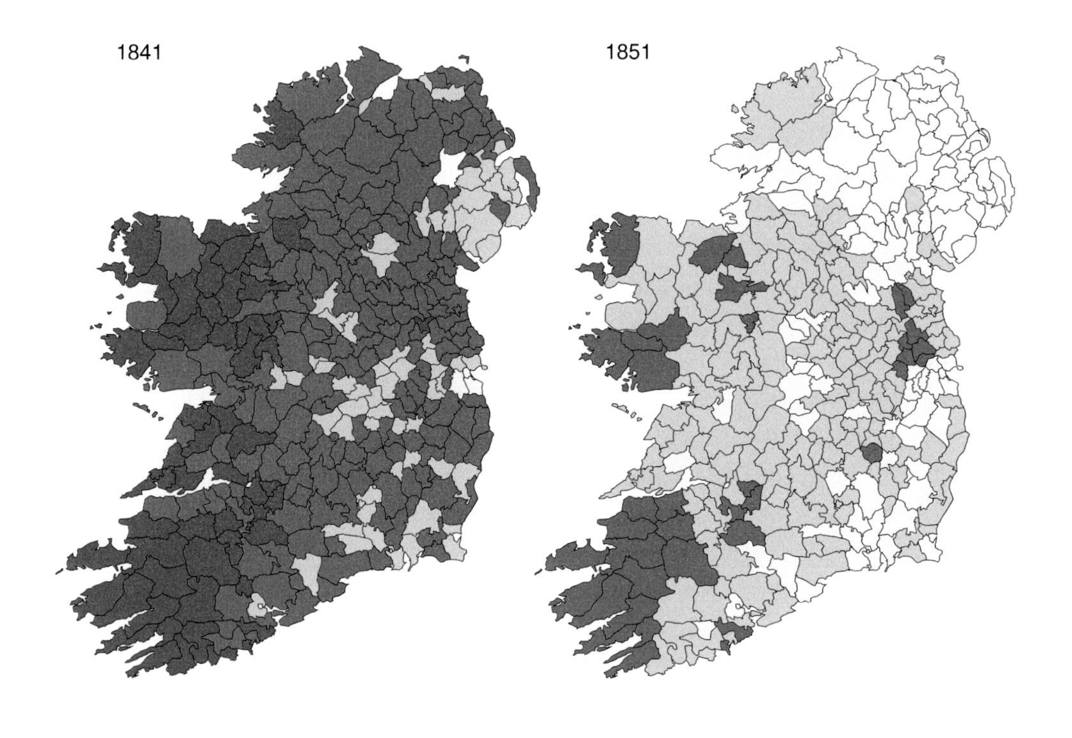

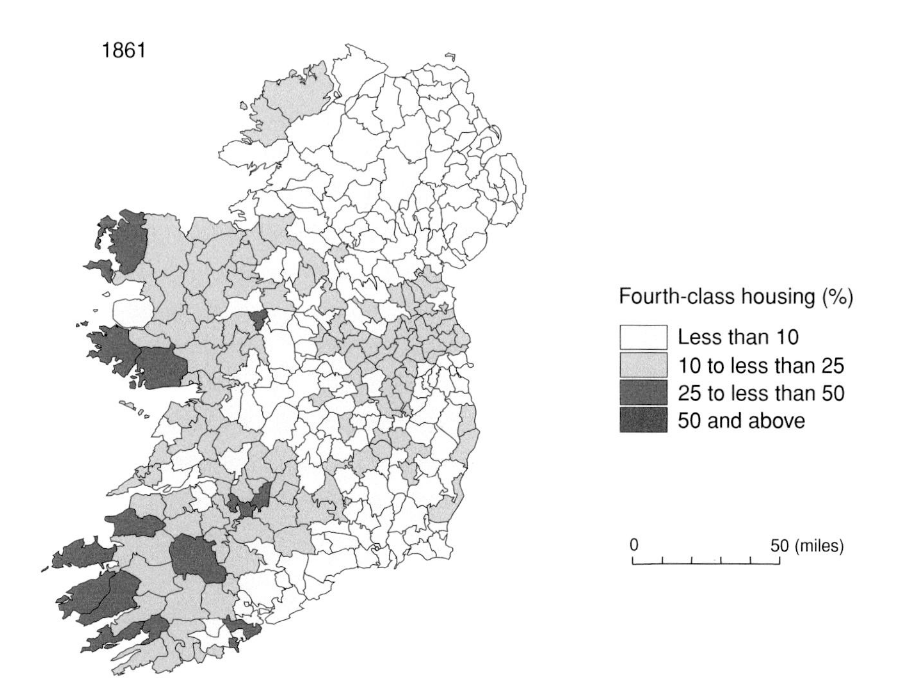

Fig. 5.4 Fourth-class housing in Ireland during the Famine years. Fourth-class housing is the lowest class recorded by the census. It consists of single-roomed dwellings made of mud. Source: Kennedy *et al.*, 1999.

consisted of one-roomed dwellings made of mud, with no windows. This attribute data has been linked to layers showing barony boundaries for the three different dates, and the maps produced.

The choropleth map is thus at the core of this type of atlas. These simple diagrammatic maps contrast with the richly artistic maps found in the Historical Atlas of Canada. Another contrast is that the GIS maps tend to be very data-led, with one or more maps for each dataset accompanied by a page of text that describes the patterns and sometimes puts them into their historical context. The Historical Atlas of Canada focuses more on historiographical themes. A third contrast is in the production quality of the two types of atlas. The Historical Atlas of Canada uses an oversized page size and is printed in full colour on high-quality paper. It retails for around £160 in the UK. The other atlases all use A4 page sizes or smaller, and only one, Woods and Shelton, uses full colour. They are available as paperbacks, and sell for around the price of a normal academic paperback. This reflects a constant problem with paper atlases that is unlikely to be resolved in the near future: to publish in a way that does justice to the maps is prohibitively expensive and is likely to seriously limit sales; however, publishing at an affordable price will undermine the quality of the publication. A final point that needs to be made on this subject is that the Historical Atlas of Canada benefited from a team of highly experienced, professional cartographers. The GIS atlases were produced by academics who were subject experts, rather than cartographers. This does lead to poorer quality maps, including some basic cartographic mistakes in choices of class intervals and shading schemes. The lesson from this is that, regardless of GIS, producing atlases with good-quality cartography remains an expensive and time-consuming business in terms of publishing costs and the cost of creating the maps and performing the required scholarship.

It is thus possible to produce cheap atlases using GIS and, even given their limitations, these can present sources in new and challenging ways. There is, however, a more fundamental question about these atlases that gets to the heart of not just using GIS to produce atlases, but to the role of mapping in GIS as a whole. All historical paper atlases follow the same general format: they split the history into a variety of themes, and put one or more maps that attempt to summarise the theme on one page, with a short length of text which describes the maps and places it in context on the facing page. The question is, to what extent does this advance our understanding of the topic? The answer is usually, unfortunately, not very much. The maps are entirely descriptive, while the accompanying text tends to be a combination of a brief background to the historiography, a description of the map pattern, and maybe a brief and tentative explanation for this pattern. To really advance our understanding of the topic, far more emphasis needs to be placed on the explanation and less on background and description. In addition, the atlas will have a large number of map plates on separate themes, each with its accompanying text. There is rarely much of an attempt to bring all of these themes together to tell the story of the topic as a whole, and how the different themes interact within this.

The risk, therefore, is that atlases tend to be an attractive but somewhat superficial product, where the maps and descriptive text are over-emphasised, while analysis and explanation are given only limited significance. This is a risk for GIS as a whole, where slick graphics, clever technology and exciting presentation are emphasised at the expense of high-quality scholarship that attempts to describe and explain the geography of the research topic.

5.5 DISTORTING MAPS: THE USE OF CARTOGRAMS

A basic problem with choropleth maps is that the urban areas in which most people live tend to be small, while rural areas with low populations tend to be large. The impact of this is that human perception tends to focus on the larger polygons when interpreting the patterns on the map. As a result, our understanding of the pattern shown may be strongly biased in favour of rural areas. This clearly has the potential to lead to a distorted view of the distribution of a variable. It can also be a problem with maps of urban areas where the enumeration districts, wards, or tracts around the edge of the city are far larger, but have smaller populations than densely populated inner city zones. It can be argued that this is a fundamental flaw with choropleth maps: the importance they give to a polygon depends on its land area, whereas to a map reader the importance of the zone is more likely to be related to its population size or some other variable.

A response to this has been to develop *cartograms*, a type of map that sacrifices locational accuracy to allow relative importance of features to be better stressed. Cartograms are not a new idea, but traditionally producing them was a lengthy task performed by a cartographer who made a large number of subjective judgements in the relative distances, positions and areas of the features shown. The advent of computer mapping through GIS has led to a reawakening of interest in this area, because large numbers of choropleth maps are now being produced, and because algorithms can be developed to produce cartograms automatically.

Dorling (1996) develops a methodology for automatically producing cartograms, where the size of a polygon is proportional to its population or some other variable. The challenge in producing cartograms is to change the size of administrative units while keeping their locations approximately the same. Ideally, the polygons on a cartogram should maintain their connectivity to the surrounding polygons, in the same way that the polygons on the original choropleth do. Dorling's methodology does not follow this idea, but, instead, converts all of the polygons into circles whose sizes are proportional to their population or other variable. The positions of the circles are then adjusted to keep them in approximately the correct place and with approximately the same neighbours, without the circles overlapping.

An example of a cartogram produced using Dorling's methodology is shown in Fig. 5.5, along with the same map as a conventional choropleth. The maps show infant mortality in the 1900s. The choropleth strongly suggests that much of the country has relatively low levels of infant mortality, and suggests that the high levels are relatively rare, although someone with knowledge of the geography of England and Wales would note that the high rates are found in urban areas. The cartogram gives a very different perspective. It clearly shows that in the 1900s, much of the population lived in urban areas, particularly London, which all but disappears from the choropleth map. As a result, the high rates of infant mortality appear far more common. In actual fact, the class intervals, which are the same on both maps, put 20 per cent of the population into each class. The class with the lowest rates covers far more of the area of the choropleth, as infant mortality was strongly concentrated in urban areas. It is interesting, however, that, on the cartogram, the dark shading for the worst rates appears to be the most common. This is a well-known cartographic illusion whereby intense shading tends to appear more important than paler shading. In summary, the choropleth shows us that the average square kilometre of England and Wales had relatively low infant mortality rates, whereas the cartogram shows us how infant mortality rates affected the average person.

Dorling (1995) publishes an entire atlas of Britain using cartograms. Other than the cartogram base layer, it is very similar to the GIS-based atlases described above. It takes a limited number of layers of spatial data, links them to a large number of variables, and maps the resulting patterns. Although it suffers from the weaknesses of the atlases described above, it is interesting in that it challenges perceptions of the demography of Britain in a visual way.

Cartograms have also been used in historical research. Gregory *et al.* (2001b) use them to help in their exploration of poverty in England and Wales through the twentieth century. Gregory (2000) uses them as part of an analysis of changing migration patterns in England and Wales. The advantage of using them in this type of research, which involves national-level change over the long term, is that the relative sizes of parts of the country change in response to population re-distribution over time. This means that processes like urbanisation and counter-urbanisation are

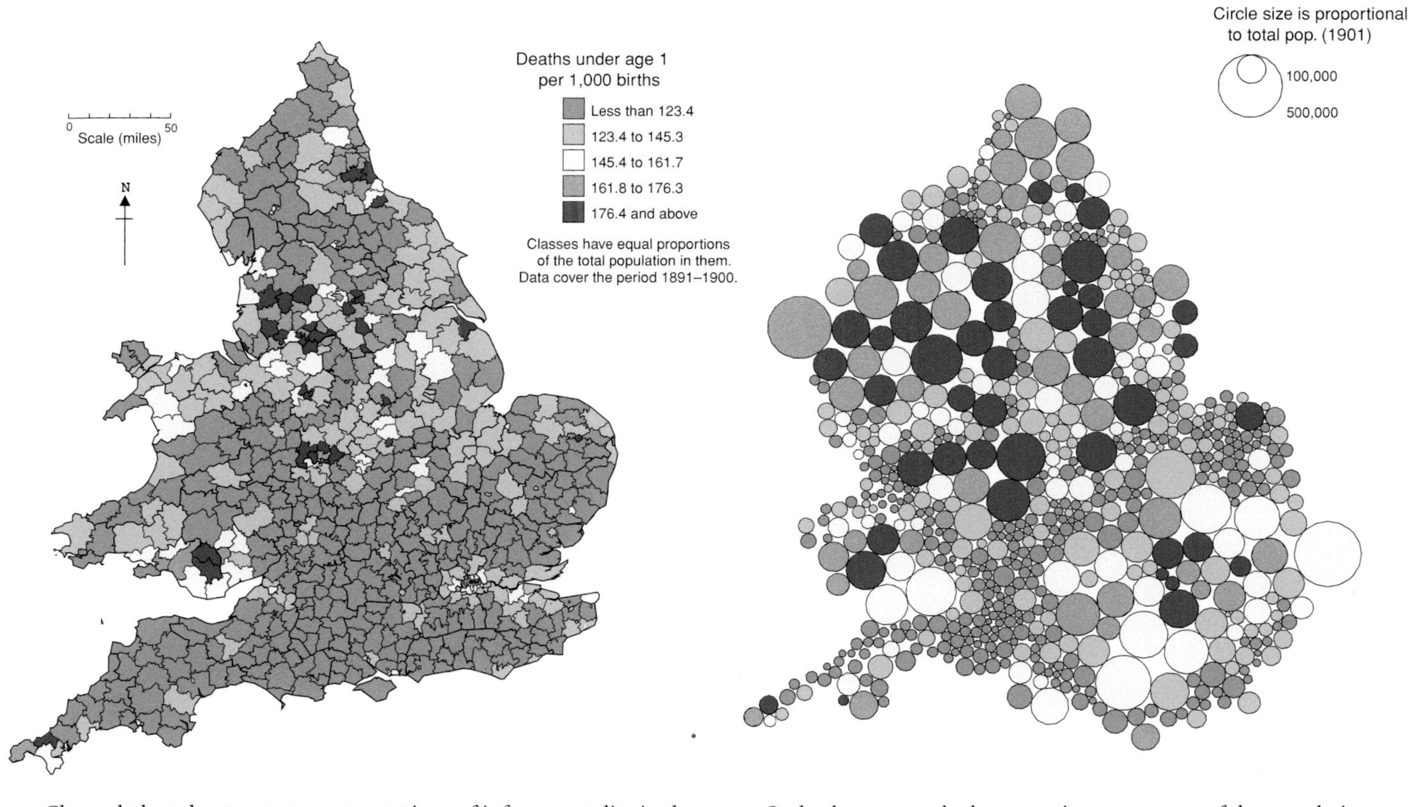

Fig. 5.5 Choropleth and cartogram representations of infant mortality in the 1900s. On both maps, each class contains 20 per cent of the population. Source: *Registrar General's Decennial Supplement*, 1911.

reflected in the cartograms, and therefore a more accurate visual impression of where the population lived at different times, and how this changed, can be developed.

One difficulty with this type of cartogram is that they distort locations to an extent that it can be difficult for the map reader to know where they are looking at. Another is that they are unfamiliar representations of the country, and they may not gain user acceptance. A solution to this is to place the cartogram alongside a conventional choropleth, as in Fig. 5.5, but this does not entirely solve the problem, and it takes up a large amount of page area, which can lead to its own problems with cartography, map interpretation and publication.

A more recent cartogram-generating methodology moves from proportional circles to polygons that preserve the connectivity between each administrative unit and its neighbours (Gastner and Newman, 2004). This gives a representation of the data that is, perhaps, more appealing than Dorling's approach, although whether it offers more insight into the underlying patterns in data is open to question. As yet, no historical research has used this approach. It received widespread attention after the 2004 US presidential election, where it was shown that the cartogram made the size of the Bush victory appear smaller than it was on the conventional choropleth. This is because the Republican vote was strongest in the larger rural states in the south and west, whereas the Democrats tended to win the smaller but more densely populated states along the coasts (Gastner *et al.*, 2004).

A cartogram is, therefore, simply a variant on a choropleth map that attempts to increase the prominence of polygons with large populations at the expense of polygons with small populations. In some cases, the effect of this can be quite marked, particularly in patterns where the population living in small zones behaves quite differently to the population living in large zones. The cartogram is a potentially useful exploratory tool for describing spatial patterns. It does, however, have most of the limitations of choropleth maps in that it grossly simplifies attribute, and change over time cannot be easily handled. It is a useful tool, but no more than that.

5.6 MOVING IMAGERY: ANIMATIONS AND VIRTUAL WORLDS

The widespread use of computing has meant that it is now possible to publish maps without the limitations of paper and ink. This has led to an interest going beyond maps into moving images, such as animations – where a number of map frames are joined together to show how the pattern changes – and virtual worlds, where images of three-dimensional landscapes are produced, and the user is able to move around or fly through the landscape, and explore it from multiple views and angles.

As will be discussed in Chapter 6, GIS should be able to handle temporal information as well as information about space and attribute. As noted above, however, a problem with conventional maps is that they do not allow change over time to be handled well, as they are fundamentally snapshots of a single date, or at most a comparison of change between two snapshot dates. In theory, animations have the potential to change this by allowing a series of snapshot maps to be joined together and played as a movie, to show how the pattern changes over time. Increasingly, GIS software has the ability to create these automatically, using file formats such as animated GIFs, AVIs and MPEGs. These can be incorporated into presentations, or published on the internet or on CDs. A certain amount of enthusiasm for the use of animations to explore change over time has been generated by human geographers (for example, Dorling, 1992; MacEachren, 1994; Openshaw *et al.*, 1994; Shepherd, 1995).

The use of animations has, to date at least, been quite limited, especially in the historical field, even though this is where it might have been expected to have made a significant impact due to the interest in change over time. There is one clear exception to this: Cunfer's (2004) use of animations to explore the spatial and temporal patterns of factors associated with the dust bowl in the US Great Plains. Cunfer's hypothesis is that dust storms were, in fact, a common event before the catastrophic Dust Bowl period of the 1930s. To illustrate this, he takes information on a variety of factors, including accounts of storms from newspapers, the spread of agriculture from census, and meteorological data. He aggregates all of the information to county level, and uses animations to show which factors are and are not occurring at different times. From this, he is able to produce a compelling argument that dust storms occurred well before the 1930s, and were found in areas that were not extensively cultivated. (For more on Cunfer's other work on this subject, see Chapter 9.) Unfortunately, while he was able to present the animations at conferences in a highly effective way, once they are published on paper (Cunfer, 2002 and 2005), their impact is somewhat diminished by having to resort to a small number of static snapshots.

Cunfer's animations are relatively simple. The spatial data consists of 280 counties which are rectangular, of similar sizes, and do not change much over time. The attribute data is also quite simple in that it consists, in some cases, of nominal data, such as whether or not there was a storm in a county in a year, or, in some cases, of hierarchies of, for example, temperature. This means that the patterns in the data are relatively easy to observe through space, attribute and time. These show that dust storms were more common than conventionally believed in the fifty years prior to the 1930s, and that they seem more related to drought than to over-intensive agriculture.

There are few, if any, other effective uses of animations in historical research to date. There are a variety of reasons for this, the most fundamental of which is cartographic. As stated above, the challenge for all thematic maps is to simplify the data without over-simplifying them. This usually involves extensively simplifying attribute. Even once this is done, the spatial patterns may still be hard to discern without effective cartography because human perception is surprisingly limited in what it can distinguish. By adding the temporal element and making the map change at regular intervals, the demands on human perception are greatly increased, compared to conventional maps. This means that when time is added, the other two components of the data – space and attribute – need to be even simpler than they are on a conventional map. Cunfer is able to do this by having relatively simple spatial and attribute data and two clear temporal questions: 'how were dust storms distributed over space and time?' and 'how did this compare with other variables?' With more complex data, this becomes more problematic.

This is not to say that there is only a limited future for animations; merely that their cartography needs to be better understood, and requires further research. Just as a thematic map becomes an effective tool for exploring spatial variation by simplifying attribute, an animation can become an effective tool for showing change over time, but to do so requires simple spatial and attribute data. Thus, the animation becomes an additional tool to the static map, but does not replace it.

One final point about animations is that they need not be used to show temporal change. It is possible to animate snapshots according to theme – for example, to show, perhaps, the different proportion of the vote achieved by each party in a general election, or the location of different employment classes. Thus the animation is only concerned with a single date, but each frame shows the proportion of the vote taken by each party, or the employment in each sector, and allows this to be compared through the animation.

Chapter 2 introduced the concept of a digital terrain model (DTM) which can be used to create a three-dimensional impression of the landscape. These can be used to create *virtual landscapes*, representations of the Earth's surface that are far less abstract than conventional maps. They attempt to present an image of what the landscape looks like, or looked like in the past, from the point of view of an observer standing on or above the Earth's surface. This involves taking the shape of the landscape, the hills and valleys, from a terrain model, and adding other features, such as the rivers, transport routes, vegetation and settlements, from other layers of data, including, perhaps, areal photographs and satellite images (see Camara and Raper, 1999; Raper, 2001; Fisher and Unwin, 2002).

These have been used in historical geography to re-create prehistoric landscapes in an attempt to understand how they would have appeared to people in the past. A good example of this is presented by Harris (2002), who uses the technology to explore the Grave Creek Mound near Moundsville in West Virginia. This is a burial mound built by the Adena people around 2,500 BP. The mound is found on the Ohio River valley in an area that, today, is heavily urbanised and industrialised. Harris's aim is to re-create the landscape around the mound as it would have appeared to the Adena people. This involved creating a terrain model of the landscape and then trying to recreate the vegetation based on sources such as pollen records. This is, to a certain extent, a work of art, as it is very much painting a picture of how the landscape is believed to have appeared. With this, he is able to create movies of what the view from the top of the mound would have been like, how the mound would have appeared to people approaching it, and so on (see Harris *et al.*, 2005). They are also able to use techniques such as hill-shading to add shadows, to see what the landscape would have been like at different times of day, and removing leaves from the trees to model the changing seasons.

As well as being a scholarly tool, virtual worlds can also be an effective communication device. A good example of this was a Granada Television programme's use of a terrain model to show the layout of the armies at the Battle of Hastings in England in 1066. Figure 5.6 shows an image used in this research. It uses a DTM to show the shape of the landscape and to allow the user to look down on the battle site from any angle. The DTM uses a large degree of vertical exaggeration to emphasise the importance of the topography. A modern areal photograph has been draped over the image to provide context. Finally, a layer of polygons has been added, to show the approximate locations of the two armies. This provided an effective demonstration that King Harold's army held the high ground, and William was at a severe disadvantage, as he had to attack up the slope (Schaefer, 2003). It is important to note that there is a significant amount of vertical exaggeration used, so the difficulties of attacking up the slope may be over-stated. Also, the precise locations of the armies at the battle are not well known; indeed, the site of the battle itself is a subject of some speculation. Therefore, while this presents a compelling image of the battle site, the user has to be somewhat cautious about how it is interpreted.

The use of virtual worlds has yet to prove itself by making an important contribution to historical research; however, it is a field that is in its early stages and does have clear potential.

A major problem with both animations and virtual worlds is that they can only be published electronically, and therefore continue to have a relatively limited impact

Fig. 5.6 The Battle of Hasting, 1066. This image uses a DTM to provide the shape of the landscape. A modern areal photograph has been draped over it, and polygons have been digitised to show the approximate locations of the two armies. Source: Schaefer, 2003.

at present, while paper publication remains the prime method of academic output. A second problem is the temptation to be overly impressed with the graphics, and forget the scholarly message that the image, be it an animation or a virtual world, is trying to create. This is a temptation for both the creator and the readers. When looking at such images, the question for the observer must be 'what does this tell me?', not 'does this look good?' If the answer to these questions are that it looks good but provides no information on the subject, the image provides no useful purpose. This may seem an obvious statement, but it is very easy to be impressed with high-quality graphics when the researcher should actually be critical of the image and of the message that it is attempting to portray.

5.7 ELECTRONIC PUBLICATION AND GIS

As GIS uses large databases from which a wide variety of high-quality graphical outputs can be created, paper publication – typically in grey-scale with limited space and no potential for user interaction – seems a very limited medium with which to publish. The growth of GIS has run in parallel with rapid developments in other aspects of computing – in particular, the ability to cheaply produce CD-ROMs that contain large amounts of data, and the development of the internet. In theory, these developments mean that the traditional constraints to publishing have disappeared, as almost everything can now be published electronically. In practice, however, the academic world has moved fairly slowly into electronic publishing. There are a number of reasons for this. Some of these are fairly fundamental – for example, although a computer screen offers the potential for colour, animations and other clever technologies, many people still prefer to read from a piece of paper. There are still difficulties in assigning academic credit to electronic outputs, whereas for paper products these are well developed. Finally, publishers have also moved with a certain amount of caution, which is hardly surprising given that their business models have been developed around paper publication. Eventually, however, academics and publishers will come to terms with electronic publication, although the extent to which readers will prefer to remain with books is as yet unclear. Electronic publication raises some fundamental technical and philosophical issues, and these, together with how they relate to publishing historical GIS resources, will be discussed in this section.

Publishing GIS data electronically is relatively simple. It usually involves having a software package that offers some of the basic GIS functionality, together with a number of layers of data. The software functionality usually includes the ability to map the data, to turn layers on and off, to pan and zoom, to change the cartographic elements such as the shading, to perform spatial and attribute queries, and perhaps to produce graphs and tabular summaries from the data. This is provided over the internet, using products that run within a web browser (Harder, 1998) such as ESRI's Arc Internet Map Server (also known as ArcIMS) (ESRI, 2005), although other products are available. For publishing on CD, a similar approach is used. Chapter 7 discusses the implications of putting GIS data on the internet in more depth.

Electronic publishing involves providing the user (or reader – the distinction becomes increasingly blurred) with the raw data, and allowing them to explore it for themselves. An early attempt at this approach was conducted by the Great American History Machine (Miller and Modell, 1988). This took county-level data from the US censuses and presidential election results from 1840 to 1970, and placed them in a system where the user could select the variables that they were interested in, and map and query them. Unfortunately, this project was somewhat ahead of its time and was never properly published commercially.

A more recent product was the Counting Heads project, based on SECOS software (Gatley and Ell, 2000). At the core of this is a CD-ROM with a variety of British and Irish census, Poor Law, and vital registration data from 1801 to 1871. SECOS allows the data to be mapped, and graphs, pie charts and other diagrams to be created. Shading and class intervals can be changed on the maps, and polygons can be queried to show their attributes. The attribute data can also be explored in tabular form. It is possible for the reader to develop a range of derived statistics from the data provided. Users can also import their own data into the system. While SECOS is not as powerful as full GIS software, it is easy to use and inexpensive. It allows an audience unfamiliar with GIS to explore these datasets, and is particularly marketed at schools, with a range of work packages being included. Figure 5.7 shows a screenshot from SECOS. The user has created a map of county-level census data for Ireland in 1851, shown on the right-hand side of the figure. The left-hand side shows the underlying attribute data that the map was generated from, and which can also be explored by the user.

Similar approaches can also be taken on the internet. The North American Religion Atlas (NARA) provides data on religion from the US censuses of 1970 to 1990, and allows the user to map these at county, state and regional level (Lancaster and Bodenhamer, 2002; Polis Center, 2001). This provides mapping functionality similar to the SECOS project, but in this case over the internet. In addition, NARA includes a variety of texts on various aspects of American religion, and thus moves more towards a conventional atlas than, for example, SECOS. The Social Explorer website provides similar functionality to access US census data from 1940 onwards. It has a very easy-to-use interface, and allows the user to develop animations of snapshots of the data (Social Explorer, 2006).

The Vision of Britain through Time project (Southall, 2006; Vision of Britain, 2005) takes a slightly different approach. This is based on the Great Britain Historical GIS (Gregory *et al.*, 2002; see also Chapter 9), but rather than encouraging the user to create national-level maps, it focuses on allowing the user to explore localities as they have changed over time. This is done by first prompting the user to enter the postcode or name of the place they are interested in. The site then allows them to explore time-series graphs of how this place, or more precisely district, has changed over the nineteenth and twentieth centuries, by drawing on a range of census, vital registration and related statistics. Choropleth maps of the data can also be drawn. The site also includes a number of historical maps as raster scans, a wealth of information on the history of administrative districts, and descriptions of places drawn from historical gazetteers and the accounts of people who have travelled through the area at times in the past.

At a very different scale, Lilley *et al.* (2005b and 2005c) produce an electronic atlas of the new towns of medieval Wales and England. This provides detailed maps of thirteen towns established by Edward I, as they would have appeared around 1300. These have been reconstructed from a variety of sources, including modern and historical mapping, field surveys, archaeology and documentary sources. The user is

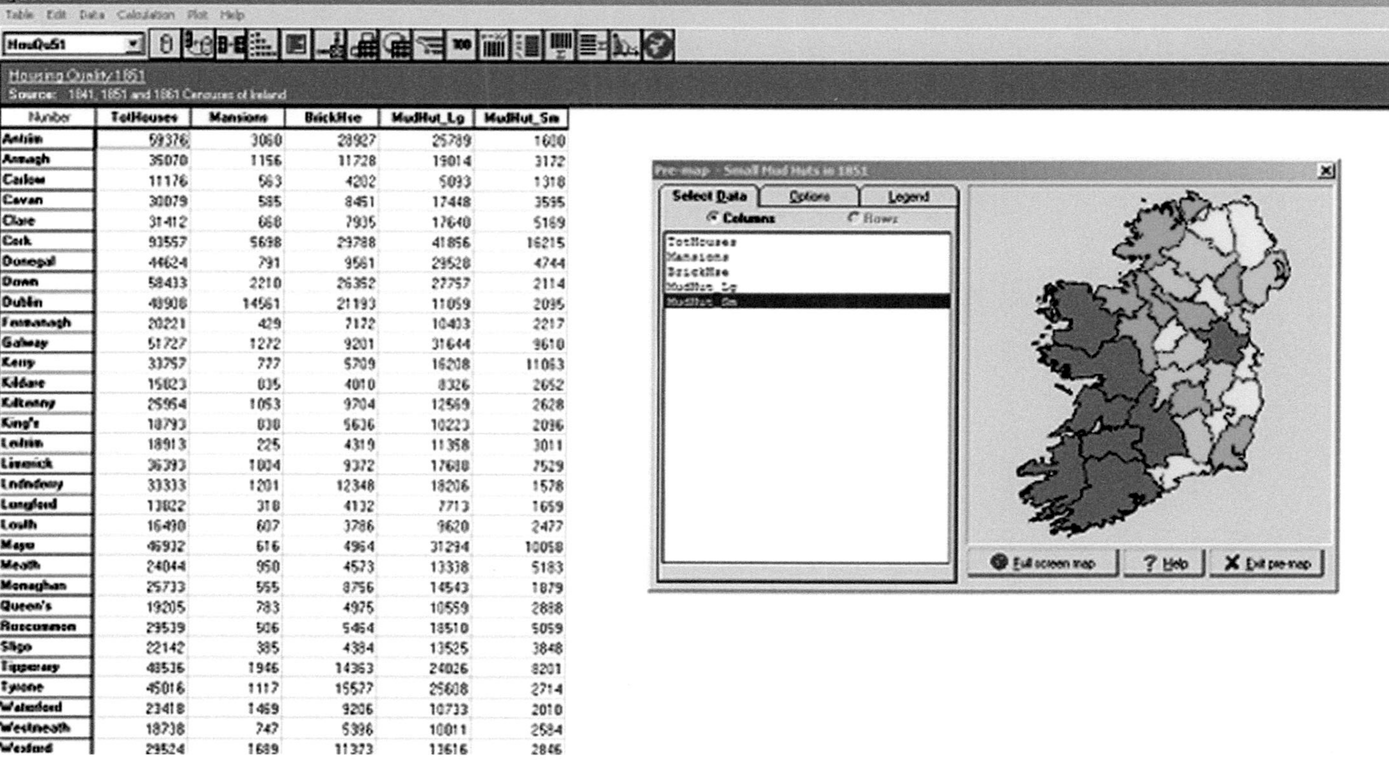

Number	TotHouses	Mansions	BrickHse	MudHut_Lg	MudHut_Sm
Antrim	59376	3060	28927	25789	1600
Armagh	35070	1156	11728	19014	3172
Carlow	11176	563	4202	5093	1318
Cavan	30079	585	8451	17448	3595
Clare	31412	668	7935	17640	5169
Cork	93557	5698	29788	41856	16215
Donegal	44624	791	9561	29528	4744
Down	58433	2210	26352	27757	2114
Dublin	48908	14561	21193	11059	2095
Fermanagh	20221	429	7172	10403	2217
Galway	51727	1272	9201	31644	9610
Kerry	33757	777	5709	16208	11063
Kildare	15023	835	4010	8326	2652
Kilkenny	25954	1053	9704	12569	2628
King's	18793	838	5636	10223	2096
Leitrim	18913	225	4319	11358	3011
Limerick	36393	1004	9372	17688	7529
Londonderry	33333	1201	12348	18206	1578
Longford	13822	318	4132	7713	1659
Louth	16490	607	3786	9620	2477
Mayo	46932	616	4964	31294	10058
Meath	24044	950	4573	13338	5183
Monaghan	25733	555	8756	14543	1879
Queen's	19205	783	4975	10559	2888
Roscommon	29539	506	5464	18510	5059
Sligo	22142	385	4384	13525	3848
Tipperary	48536	1946	14363	24026	8201
Tyrone	45016	1117	15577	25608	2714
Waterford	23418	1469	9206	10733	2010
Westmeath	18738	747	5396	10011	2584
Wexford	29524	1689	11373	13616	2846

Fig. 5.7 A screenshot from the SECOS program. The map on the right has been created from the attribute data on the left.

able to zoom in and pan around the maps, turn on layers of other information taken from the more recent maps, and query certain features.

Although all of these products are forms of electronic publication, they raise some interesting questions about the nature of publication itself. A conventional academic publication takes a dataset or source and analyses it. The reader is then presented with a *fait accompli* argument in which the writer presents what he or she believes that the source tells them. The form of electronic publication presented by the works described above is entirely different. They take a dataset, or number of datasets, and publish them using maps and diagrams. Little or no attempt at interpretation is made. The user (rather than reader) is then left to conduct their own analysis and draw their own conclusions.

This raises the question, to which there is yet no clear answer, as to whether these are academic publications at all. On the one hand, there is a tradition in history of researching a topic in depth and publishing a resource, based on the research, that offers no interpretation. Darby and Versey's (1975) *Domesday Gazetteer* is an obvious example. The resources described above all go beyond mere transcriptions. At the very least, they join statistical data to administrative boundaries, but in most cases they integrate a large amount of data from different sources, and include some descriptive text. They are, therefore, well removed from the raw data and have usually involved a large amount of painstaking academic research. None of the products, however, offer much, if any, interpretation, and none present anything that resembles a conventional academic paper or book. There are no standards, at the moment, for peer reviewing such products. This has some quite important implications. The fact that the *Domesday Gazetteer* was published by a well-known and respected academic press, and had been through peer review, means that readers are able to place some confidence in it, and the authors were given academic credit for their work. With resources found on the internet, there is little to guide the reader as to how good the quality of the product is, and no mechanisms for giving credit to the authors. This is not to cast aspersions on any of the products described above; however, with the growth of the internet and the ease of publishing on it, this is a serious issue that the academic community needs to address.

The alternative route that electronic publication can follow is to present something similar to a traditional academic paper, in that it tells a story based on sources and data, but to incorporate far more of the source material into the publication. This allows the person using the product to approach it both as a reader and as a user. As a reader, they can follow the argument presented to them in the usual way. As a user, they can explore the underlying data in ways that would be impossible in a conventional paper publication. The best developed example of this in historical research is Thomas and Ayers's (2003) publication of the Valley of the Shadow project (described in more detail in Chapter 9). This is published in the *American Historical Review*; however, the paper publication merely provides a description of the choices the authors made in creating the electronic publication. This is available from www.historycooperative.org/ahr/elec-projects.html,[1] and in turn is based on the Valley of the Shadow archive, available at valley.vcdh.virginia.edu.[2] The electronic article is similar to a structured paper publication, in that the reader can read through the argument, following from section to section. Where it differs is that the reader can also follow through topics in considerably more depth, by exploring maps, examining original texts, and studying raw datasets. This is, therefore, a more in-depth product than a conventional paper publication, as the reader is not only presented with the authors' arguments but is also given access to the sources that the authors based their arguments on. There is also an assurance of quality, as the paper has been peer reviewed by an established journal.

There are, therefore, two distinctly different approaches to electronic publication in historical GIS. The first involves publishing data and allowing the user to explore it themselves; the second involves publishing an argument, but including the data that the argument was based on. Both approaches have advantages and merit, and it may well be that the gap between these two approaches narrows over time. There are still issues to be resolved, such as peer review of datasets and resources; however, this is bound to be an area of considerable growth, and one in which historical GIS must be closely involved if it is to make best use of its large and complex datasets and the potential for high-quality graphical output.

5.8 CONCLUSIONS: MAPS AND STORIES

The map is often the best way of communicating the information held by a GIS database, both to the researcher during the research process and to the reader once the research is finished. The forms of maps produced by GIS are very simple. Typically, they are thematic maps of a single layer of data for a single point in time, whose attribute has been simplified to make the patterns clearer to the reader. Producing effective maps is not as easy as GIS software makes it appear. Cartography is a complex and subjective skill, whose aim is to communicate information effectively through the limits of the media used and of human perception. It is not something that can be done effectively by accepting the defaults on a software wizard.

[1] Viewed on 18 June 2007. [2] Viewed on 18 June 2007.

Even if the maps produced are effective, they are still highly limited. It is interesting to consider how much one learns from historical atlases, whether they are produced using GIS or traditional means. After careful reflection on this question, we believe that the answer to this is that we learn little from these. Instead, they provide an attractive but simplistic understanding of the historical geography in question. This has lessons for GIS. The problem with atlases is that the maps tend to be over-emphasised at the expense of the more detailed argument that is best presented in text. Clearly, this is a real danger with GIS, where maps can be produced almost *ad infinitum*. This leads to the risk that GIS will simply be used as a tool for crude spatial-pattern spotting. To prove its worth as a scholarly tool, researchers using GIS must go beyond this, to ask 'what lies behind the patterns that the maps show?' and to describe and explain the historical processes that form these patterns.

Advances in technology open up new potentials to visualise and publish GIS data. These are currently in their early stages, but they offer exciting new potential and many challenges for the scholarly community, not least how to peer review these products. We must always be aware, however, that the aim of this technology should be to provide the reader or user with spatial information in an understandable way that enables them to explore scholarly questions. Their utility in doing this must be critically evaluated, rather than simply being impressed by fancy 'Nintendo geography' graphics. Nevertheless, animations, virtual worlds and electronic publishing open exciting and entirely new ways of conducting historical scholarship that no previous technology has allowed.

GIS databases are time-consuming and expensive to produce. Thematic maps are, perhaps, better thought of as diagrams, rather than maps. When these two statements are considered together, it is clear that investing large amounts of time and money simply to produce a large number of diagrams is unlikely to make the investment worthwhile. The maps that GIS produces are an effective way for a researcher to explore the spatial patterns in his or her database, and of presenting an argument to their readers. They do not, of themselves, provide answers to any questions beyond 'what are the spatial patterns in this dataset?' Maps, therefore, are effective at illustrating a story, but do not tell it. Telling it requires detailed and thorough scholarship that emphasises both the historical and the geographical patterns within the data.

6

Time in historical GIS databases

6.1 INTRODUCTION

GIS has the ability to handle thematic (or attribute) information and spatial information to answer questions about *what* and *where.* Most information also has a temporal component that tells us *when* an event occurred, or when a dataset was produced. This component of information is not explicitly incorporated into GIS software, and has attracted only limited interest among GI scientists. A lack of temporal functionality in GIS software is commonly criticised. Progress in adding it has been limited, as GIS software vendors do not see this as particularly important for their market. As a result, researchers wanting to handle temporal data are largely left to make their own decisions about how they are going to do this within GIS.

Chapter 1 introduced Langran and Chrisman's (1988) idea that, faced with the complexity of data with thematic, spatial and temporal components, the traditional approach has been to fix one component, control the second and only measure the third accurately (see also Langran, 1992). They give the example of census data, where time is fixed by taking the census on a single night, and space is controlled by subdividing the country into pre-defined administrative units. The theme is the number of people counted in each unit, which is well handled at the expense of the other two components. Soils mapping gives a slightly different example. Again, time is fixed, but in this instance theme is controlled by subdividing soils into classes or types, and space is measured accurately, as the map shows the boundaries between the different classes. Table 1.1 presents more examples of this. Ideally, there would not be this limitation. All three components of the data should be handled simultaneously and in detail, subject only to the limitations of the source. As we have seen, GIS allows this to be done with attribute and space. In this chapter, we explore its limitations with the temporal component, and what can be done to overcome these. It is important to stress that this is a scholarly issue, not solely a technological one. As was discussed in Chapter 1, geography is the study of places and the interactions between them. History is the study of periods of time. Temporal change may be an important part of understanding this, but it does not have to be (Baker, 2003). This leaves the question, 'how does the ability to improve our handling of spatio-temporal data assist in our understanding of historical geography?' We are some way from answering this, but parts of this chapter will explore how the complexity of spatio-temporal data has traditionally undermined our understanding of change over time and space, and address how historical GIS may help to resolve this.

Section 6.2 explores why space and time are important to historical geographers, and the extent to which the inability to cope with them in an integrated manner has limited our understanding of the discipline. Section 6.3 explores the strengths and limitations of how time is handled in GIS databases at present. The last two major sections explore how GIS can be used to help handle spatio-temporal information: section 6.4 looks at database architecture that can hold spatio-temporal data, looking in particular at national historical GISs, where most research to date had been focused; and section 6.5 looks at how GIS functionality can be used to manipulate data published for incompatible administrative units directly, so that they can be made directly comparable over time.

This chapter focuses on database issues, and describes architectures that can be used to answer queries such as 'what was at this place at this time?' and 'how has this place changed over time?' Ways of visualising change over time using animations where discussed in Chapter 5.

6.2 SPACE AND TIME IN HISTORICAL GEOGRAPHY

Historical geographers and, indeed, human geographers more generally have a long tradition of arguing for the importance of including time into their research (see, for example, Butlin, 1993 and Dodgshon, 1998). The problem is that, while there are good justifications for including time, to date, much of the actual research in this area has suffered due to the difficulties of handling data that contains thematic, spatial and temporal information. This section briefly discusses the arguments for including time, and the reasons why this has been less effective than might be hoped.

In the early 1970s, Langton (1972), arguing for the use of systems theory in human geography, identified two approaches to how change over time can be studied. He termed these *synchronic* analysis and *diachronic* analysis. *Synchronic* analysis assumes that a system starts in equilibrium, undergoes a change due to some form of stimulus, and then reaches equilibrium again. The changes in the parameters between the two equilibrium states can be compared to explore the impact of the stimulus on the system. *Diachronic* analysis attempts to take a more realistic approach by tracing the elements of a system and following their changing functions over a successive series of intervals as the system evolves. He argues that this is a much more effective and realistic way to study the way that processes act once a system has come under pressure to change, either as a result of an external stimulus or due to internal pressures.

At around the same time, Sack (1972; 1973 and 1974), talking in terms that are very relevant to GIS, argued that geography was not simply a spatial science, where understanding of a location could be based on the geometric arrangements of points and lines. Instead, developing an understanding of a place also required an understanding of the processes that formed the place. These processes are, by definition, dynamic, and also involve what he termed 'congruent substances', which are human agents and underlying structures within society. As a result of this, Sack argued, space alone can never provide an understanding of what causes patterns to form, as space, or geometry, is not of itself a causal variable.

More recently, Massey (1999 and 2005) argues that space and time together are important for human geography, and that this applies across the discipline. Her basic argument is that an understanding of change over time is crucial to allow us to understand how things develop, and how the current situation was arrived at (and this can apply to a 'current situation' at any point in the past). Time thus provides the story of how a particular place or phenomenon evolved, without which it is impossible to understand the processes driving the evolution. Space, she argues, is also important because without space there can be only one story of evolution, with the risk that this implies that this story is inevitable and will inevitably occur in the same way in different places. With space, there becomes the possibility to tell multiple stories of how places or phenomena evolve to become what they are in the present. Thus 'space could be imagined as the sphere of the existence of multiplicity, of the possibility of the existence of difference. Such a space is the sphere in which distinct stories coexist, meet up, affect each other, come into conflict or co-operate. This space is not static, not a cross-section through time; it is disrupted, active and generative' (p. 274). If this is the case, it means that the story of how multiple places developed over time becomes a sequence that says 'how we got here from there' rather than a progression that says 'how we must get *here* from there' (p. 272). Thus only by acknowledging space can it be acknowledged that different places can behave differently.

This demonstrates that over the past thirty years or so there has been a constant argument that ignoring or over-simplifying space or time leads to an impoverished understanding of a place or a phenomenon. In spite of this, Jones (2004) conducts a survey of papers published in *Transactions of the Institute of British Geographers*, *Annals of the Association of American Geographers*, the *Journal of Historical Geography* and *Progress in Human Geography* over the past fifty years. He argues that these show that the focus of human geography has increasingly emphasised the present and recent past. This is interesting, as it suggests that, while there has been a consistent argument for the importance of time in human geography, the amount of research into change over time has decreased across the discipline, including in historical geography.

What then have historical geographers made of exploring change over time and space? Taking the example of quantitative exploration of long-term demographic trends in England, it is clear that the researchers have attempted to analyse change over space and time, but have been severely hampered by the complexity of their data. As a result, they have had to either focus on a small study area or time period, or had to simplify one or more of the components of the data. An example of doing the former is Lawton's (1970) study of the population of Liverpool in the mid-nineteenth century. This is based on using census data to study occupations, migrations, and the age–sex structure of the city and its immediate surrounds. By concentrating on a small study area and time period, the author is able to use the available data at the most detailed spatial level available, and, where necessary, examine change over time for the entire century. The study is, however, restricted to only seventeen separate administrative units, and some of the analysis is hampered by boundary changes. It is, therefore, vulnerable to a criticism that it only tells the story of one place, Liverpool, and the lessons learned from this place may only be of limited relevance in other places.

In a seminal piece, Friedlander and Roshier (1965) were able to work with decennial data – the most detailed available – to calculate net migration flows for the whole of England and Wales in the period 1851 to 1951. Calculating flows was complex, and makes as much use of the information content of the source attribute data as possible. The drawback is that they were forced to work at county level, thus losing a significant amount of spatial detail. Even at this scale, they were unable to quantify flows between adjoining counties because of the problems of boundary changes. This means that the spatial, and to a lesser extent the attribute, components of the data had to be simplified. Lee (1979) faced similar problems. He was able to create detailed standardised occupation structures for men and women from 1841 to 1971. Again, however, he was forced to work at county level, and even here he needed to create two series: one for 1841 to 1911 and one for 1901 to 1971, partly because of the changing occupation classifications used in the census, but also because the census changed its definition of counties after 1911. In later work on infant mortality from

1861 to 1971, he encounters the same problem and is again forced to work at county level (Lee, 1991).

There are two difficulties with working at these high levels of aggregation, similar to American states. The first is that a large amount of the spatial detail is lost, and therefore the understanding that can be gained from the data must in turn be severely reduced. In particular, all ideas of rural and urban differentiation are lost, but it is still very tempting to base interpretations of patterns on urban and rural differences. The second difficulty is that these gross levels of aggregation lead to enormous problems with modifiable areal units (Openshaw, 1984; Openshaw and Taylor, 1979; see also Chapter 8) which makes it increasingly difficult to determine whether any patterns found are simply due to the arrangements of the administrative units used to report the data, rather than underlying patterns within the data. This is a particular problem when attempting to find statistical relationships between two or more variables.

Darby *et al.* (1979) concentrate on spatial detail in their comparison of wealth in England, as measured by three key sources: Domesday Book of 1086 and the Lay Subsidies of 1334 and 1525. The spatial framework used in this paper was based on Darby's earlier work on Domesday, started in the 1930s but not published in full until the 1970s (Darby, 1977). This had allocated almost all of the 13,000 place names given in Domesday Book to 715 sub-county-level units. These units could not be directly compared with the areas used to collect the 1334 and 1525 Lay Subsidies, so some aggregation was performed to create 610 units. By doing this, the study managed to cover a long time period, but could only fit three sources into its spatial framework. The attribute data were measures of wealth that are computationally simple. This is therefore a national-level study that uses a reasonable degree of spatial detail, but that has simple temporal and attribute data. In spite of this, it required four authors, and was heavily based on nearly a lifetime's work by one of them.

Lawton (1968) manages a national-level analysis that combines spatial and temporal detail, but whose attribute detail is weak. He performs a registration district-level analysis covering the period 1851 to 1911. This means that he has approximately 635 spatial units and seven snapshots. He was also researching net migration, but was only able to look at rates for the total population, rather than subdividing them by age and sex. This gives a very simple level of attribute detail.

A final example, the contrasts between Buckatzsch (1950) and Schofield (1965), provides a warning about attempting to include too much detail without thinking carefully about the validity of the sources. Both authors looked at long-term change in wealth distribution in England and Wales from the Middle Ages. Both analyses were county level, and used relatively simple measures of wealth. Buckatzsch used no fewer than thirty property-tax-based sources to provide snapshots covering the period 1086 to 1843. Using these, he argued that the distribution of wealth remained stable from the Middle Ages to the seventeenth century, but changed greatly in the eighteenth century. Schofield re-examined Buckatzsch's evidence, but comes to very different conclusions. His argument is that some of the tax assessments that Buckatzsch used were not directly comparable, and this went some way to explaining his results. Schofield deliberately limited the temporal extent of his study and only included sources that he felt were comparable after detailed examination. He argues that there were actually major changes in the distribution of wealth in the later Middle Ages, similar to those ascribed by Buckatzsch to the eighteenth century.

Although this discussion has been limited to only a few examples, they demonstrate that analyses of change through time and space have traditionally been severely limited, not just by the data themselves, but also by the way that researchers have had to represent the data. This has usually meant that, in national-level studies, the amount of temporal information is less than the available data can support, that the data are often aggregated spatially to county level, and that the full amount of attribute detail is not used. Researchers were restricted in what they could achieve by the volume and complexity of their data, and, in particular, the problems of boundary changes. More localised studies were able to loosen these constraints, but this approach leads to the question about how representative the local study area was. This often means that Langran and Chrisman's (1988) idea of having to fix one component of the data and control another in order to measure the third was the best that could be achieved, but even this was often optimistic.

In summary, therefore, although there have been repeated calls for space and time to be handled together in geographical research, research that has managed to do this effectively is very limited due to the complexity and volume of data that have spatial, temporal and attribute components. The challenge for GIS is to improve the handling of the data, such that an improved understanding of change over time and space can be gained from the available sources.

6.3 TIME IN GIS DATABASES

In spite of the fact that researchers have been calling for temporal GIS functionality since the early 1990s (see Al-Taha *et al.*, 1994; Langran and Chrisman, 1988; Langran, 1992; Peuquet, 1994), as yet, no commercially available GIS software fully incorporates a temporal component. This means that the main focus of this section is on the theoretical literature.

Langran (1992) identifies six functions that she argues a temporal GIS needs to be able to perform. These are: *inventory*, *updates*, *quality control*, *scheduling*, *display* and *analysis*. *Inventory* refers to the ability to store a complete description of the study area and account for changes in both the real world and the computer representation of it. To do this the data structure needs to be able to supply a complete lineage of a single feature, its evolution over time, and its state at any moment in time.

Updates involve the need to replace outdated information with current information. The outdated data need to be preserved in a form that is accessible, should the user want to re-create past states, not only of reality but also of the database's representation of reality. In other words, if the database has been updated to correct mistakes, this needs to be stored *in addition* to updates that reflect changes in the real world. Strategies to do this for changing administrative boundaries have been proposed by Wachowicz (1999) and Worboys (1998). Northern Ireland's national mapping agency, the Ordnance Survey of Northern Ireland (OSNI), has included this functionality in its base mapping. This allows its core GIS to be updated in the light of new surveys, without losing the information that existed in previous surveys (Atkins and Mitchell, 1996). As yet, however, the ability to handle updates in this form is only likely to have limited relevance to historians.

Updating is linked to the third function, which states that temporal information should assist with *quality control*, by evaluating whether new data are logically consistent with superseded data. Given that historians are unlikely to be building databases that evolve to keep up with the present, this is not discussed in any detail here.

Scheduling is more concerned with predicting the future. It involves developing functionality that allows the system to identify and anticipate threshold states in the database that trigger pre-defined responses. An example of this might be by incorporating information on rates of streetlight bulb failures to inform managers of the need for action. Again, this is unlikely to be relevant to historians.

Display involves generating static or dynamic maps or tabular summaries of change over time in the study area. This was discussed in Chapter 5.

Finally, *analysis* is concerned with allowing the GIS to be able to explain, exploit or forecast the components contained by the study area, to allow an understanding to be gained of the processes at work. Doing this involves statistical analyses of trends and patterns, and creating models of the data. This is discussed in Chapter 8.

Therefore, although time in GIS can be a very complex subject, the lack of a constantly evolving present in historical databases means that updating, scheduling, and most issues to do with on-going quality control are not required. This section focuses on inventory: the ability to store a complete representation of the study area and how it changes over time. It shows that effective ways of handling time can be developed using the currently available GIS functionality.

Peuquet (1994) argues that a temporal GIS database should be able to answer three types of queries:

1. changes to an object, such as 'has the object moved in the last two years?', 'where was the object two years ago?' or 'how has the object changed over the past five years?';
2. changes in the object's spatial distribution, such as 'what areas of agricultural land-use in 1/1/1980 had changed to urban by 31/12/1989?', 'did any land-use changes occur in this drainage basin between 1/1/1980 and 31/12/1989?' and 'what was the distribution of commercial land-use fifteen years ago?';
3. changes in the temporal relationships among multiple geographical phenomena, such as 'which areas experienced a landslide within one week of a major storm event?' and 'which areas lying within half a mile of the new bypass have changed from agricultural land use since the bypass was completed?'

At around the same time, Langran (1992) presents a similar list when she argues that a temporal GIS should be able to respond to queries such as 'where and when did change occur?', 'what types of changes have occurred?', 'what is the rate of change?' and 'what is the periodicity of change?'

It is worth noting that all of these queries can be answered by conventional GIS approaches that have no concept of time; however, in many cases this would rely on long-winded techniques. Take, for example, Peuquet's 'which areas lying within half a mile of the new bypass have changed from agricultural land use since the bypass was completed?' query. This could be answered by first overlaying a layer of land use from before the bypass was created, with a layer from after it was created, to find out which areas had changed from agricultural land use. This could then be overlaid with a half-mile buffer around the bypass to find which areas lay within this buffer, which would provide the answer. Ideally, there would only be one layer of data on *changing* land-use data, and one layer of roads data that includes when they were opened. Within this structure, a single overlay operation could be used to establish what areas had changed from agricultural when the bypass was opened. Therefore, a more continuous and integrated representation of time would explicitly address when the bypass was opened and when each field's land use changed.

Langran (1992) describes three possible models that a spatio-temporal GIS database could use. The first, *time-slice snapshots*, only records the situation at certain times, and is demonstrated for four snapshots in Fig. 6.1, where each snapshot is a separate layer. Time-slices can either be taken at regular intervals or in response to

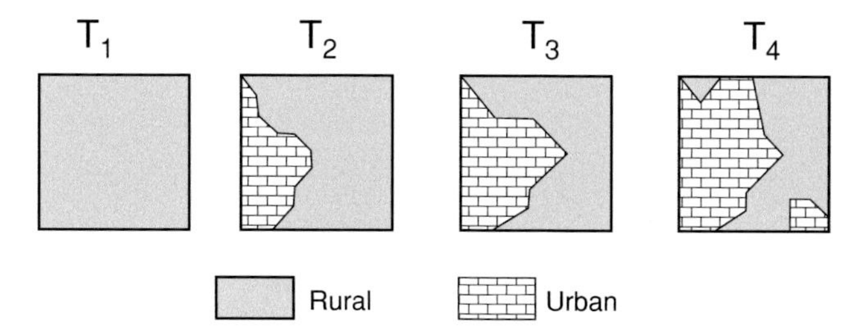

Fig. 6.1 Time-slice snapshots representing urban expansion into a rural area. Source: Langran, 1992: p. 39.

changes. This solution is intuitive, but limited for a number of reasons. Firstly, it will answer queries such as 'what exists at time T_i?', but not 'how has this changed from T_{i-1}?' or 'what is the frequency of change?' Although states are represented at snapshots in time, the events that change states are not. This makes it difficult to trap logical errors. Each layer has spatial topology in the usual way, as was described in Chapter 2. There is, however, no temporal topology because no information is stored describing how the features on one snapshot connect to features on other snapshots. As there is no concept of temporal structure, rules to enforce logical integrity are difficult to devise. Finally, a complete snapshot is produced at every date. This means that all unchanged data are duplicated, resulting in large amounts of redundant storage. This problem will be made worse if data that have not changed are taken from different sources, as digitising errors will inevitably lead to sliver polygons if any overlays between dates are undertaken (see Chapters 3 and 4). Nevertheless, this architecture may well be an accurate representation of the way that the data are available to the historian. For example, if censuses are to be used to examine change over time, the sources will be in this form. If preserving the integrity of the original source is a key aim, then using time-slice snapshots may be an effective solution.

The second model Langran suggests is what she terms *base map with overlays*. This involves standing on a timeline and looking into the past or the future. A base layer at T_0 is used to define the data's original state. At appropriate intervals, changes that have occurred since the previous update are recorded. The intervals used do not have to be regular. This is shown in Fig. 6.2. This model allows queries on both states and versions. To answer the query 'what was the data at state T_i?', T_0 is merged with overlays T_1 to T_i. To answer 'what has changed between T_i and T_j?' all overlays between T_{i+1} and T_j are merged. To answer 'what versions has this object and when did it mutate?', each overlay is checked for amendments in the object's location, and finally, to calculate the frequency of change, the numbers of mutations at each location can be calculated. According to Langran, in this model the temporal structure is

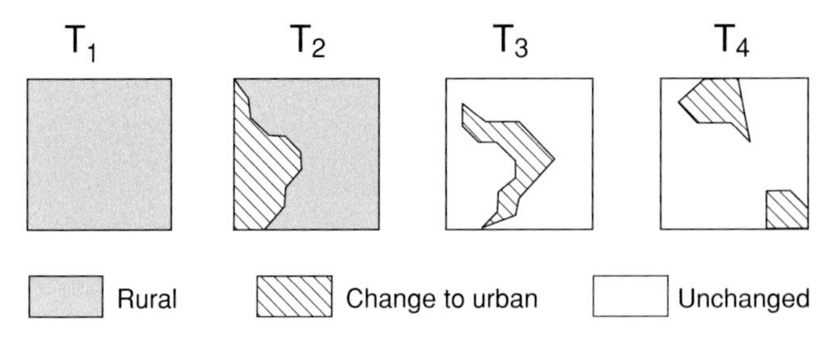

Fig. 6.2 Base map with overlays to describe the urban expansion. Source: Langran, 1992: p. 40.

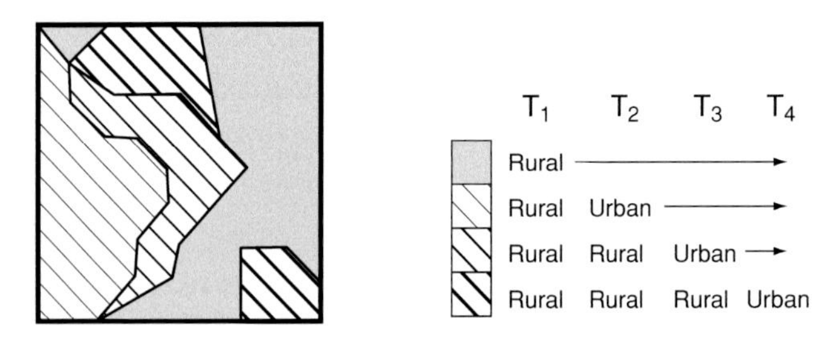

Fig. 6.3 A space-time composite of urban expansion. Each polygon has an attribute history distinct from that of its neighbours. Source: Langran, 1992: p. 41.

evident, errors can be trapped, and redundancy is minimal. This architecture would be well suited to situations where a record of updates is available, following on from a snapshot of all the data at a point in time. In many cases, constructing this architecture may involve taking data from several sources and integrating it.

The final model put forward by Langran is what she calls the *space-time composite*, shown in Fig. 6.3. This is a variation on the base map with overlays theme, but the base map becomes a temporal composite built from accumulated geometric changes. This allows units with coherent histories to be identified, and their mutations can be described by the attribute data. This reduces the components of the data from three to two, by allowing space to be treated atemporally, and time to be treated aspatially. To create a layer for time, T_i polygons are dissolved (see Chapter 4) based on the attributes at that date. Querying which areas have changed over time becomes an attribute query that asks 'which areas at T_i had changed by T_j?' The results of this query can be mapped.

A different three-way approach to building spatio-temporal databases is provided by Vrana (1989 and 1990). Working from the point of view of encoding cadastral information, he terms his approaches: *date stamping*, *transaction logs*, and *updating procedures*. With *date stamping*, every feature has its start and end dates explicitly encoded as attributes. This allows the object's location to be determined in time as well as space, but does not provide any reference to previous or subsequent states. The second approach, *transaction logs*, involves a series of events being stored to allow a chain of events to be reconstructed. This is similar to Langran's *base map with overlays* approach. Vrana, however, notes that reconstructing a composite picture based on this model can be very laborious, as a lot of redundant events may have to be assimilated. Vrana's *updating procedures* approach is based on storing the obsolete records when records are updated. This allows updates that correct mistakes to be stored, in addition to records that are updated to store change. The two latter approaches, he argues, can be useful to reconstruct the basis of historical decisions.

A variety of other researchers have also described and critiqued similar data structures (see, for example, Kampke, 1994; Raafat *et al.*, 1994; Robinson and Zubrow, 1997), but little progress has been made beyond this basic typology, except in the field of object orientation. This, it is argued, provides an alternative approach to modelling space and time simultaneously (see, for example, Peuquet, 1999; Wachowicz, 1999; Worboys, 1998). As object orientation has not yet been widely accepted amongst GIS users, this will not be described in any detail here.

6.4 APPROACHES TO HANDLING TIME IN HISTORICAL GIS DATABASES

The need to generate databases that incorporate change over time has been an important part of the development of historical GIS. While no-one has managed to fully implement all of the desirable characteristics of a temporal GIS described above, the lack of temporal functionality within commercial software has not proved a major hindrance. Two distinct situations can be identified. The first situation is changes to polygon nets such as administrative boundaries, which is the most complex situation, but also the one in which most progress has been made through the development of national historical GISs. The second situation, changes to stand-alone features such as points or individual polygons without neighbours, such as the extent of an urban area as it changes over time, is easier to manage, but has received less attention in the historical GIS literature.

6.4.1 *National historical GISs*

One of the earliest attempts at implementing a temporal GIS came from historical research. This was an attempt to create a GIS of changing American county boundaries from the 1790s to the 1970s for the US County Atlases Project (Long, 1994). A relational architecture was developed that allowed counties to be made up from boundaries in much the same way as conventional polygons are constructed. However, the boundaries had date stamps as part of their attributes, to allow them to start and end in response to boundary changes (Basoglu and Morrison, 1978). Unfortunately, this attempt occurred in the mid-1970s at a stage when GIS was still in its infancy and computing power was limited. As a result, the County Atlases Project abandoned using GIS in favour of paper-based cartographic approaches for over twenty years.

Although unsuccessful, this work was in many ways the precursor to many national historical GISs that had to solve the same problem: namely, how to store the development of changing administrative units in a GIS. A large number of countries are working on this problem, using a variety of approaches (Gregory, 2002a; Knowles, 2005b). The simplest approach is to use a version of Langran's time-slice snapshots to capture data at dates which are either seen as important examples of the arrangements of boundaries or, more pragmatically, for which source maps are available to digitise from. This is known as the *key dates* approach. A simple example of this approach is taken by Kennedy *et al.* (1999) in their atlas of the Great Irish Famine (see also Chapter 5). The atlas uses census data to show demographic changes resulting from the Famine. At its core are layers representing the different administrative geographies used to publish the censuses of 1841, '51, '61 and '71. These layers are linked to a wide variety of census data from these dates, allowing sequences of maps to be produced showing, for example, how the spatial distribution of housing conditions and use of the Irish language change over the period.

While this approach is simple and effective, it is only suitable for a limited number of dates, or, where change occurs, at clearly defined times between periods of relative stability. More complex situations are more problematic. If, for example, a researcher wanted to create a database of changing administrative boundaries for an entire country, the key dates solution would be to digitise the boundaries for every date at which maps are available. There are two problems with this: firstly, where boundaries do not change, the same line has to be digitised many times, resulting in wasted effort, redundant storage, sliver polygons, and potentially being unable to tell whether two representations of a boundary are different because of a formal boundary change or due to mapping or digitising error; secondly, boundaries can only be digitised for dates where source maps are available, and this may not coincide with the dates for which there are attribute data. Linking attribute data to spatial data for a nearby date may provide a good approximation of the actual boundaries, but there will be some error introduced as a result. This can range from an incorrect representation of the administrative unit concerned, to either polygons with no attribute data or attribute data with no polygons. Therefore, this is a pragmatic solution for a relatively short and well-mapped period of time, but is far from ideal.

A more complex version of the key-dates approach is provided by the US National Historical GIS. This system required attribute data for the US censuses from 1790 to 2000 to be linked to the boundaries of census tracts, counties and states. The project benefited from the availability of high-quality boundary files for 1990 and 2000. The system started with these boundaries, and worked backwards through time researching and recording inter-censal boundary changes. Separate layers of data were developed for each census year. The fact that changes are researched and added to a base set of boundaries means that the multiple layers formed represent an integrated history of the development of the various administrative units. For tracts, a linked attribute database describes each tract's history (Fitch and Ruggles, 2003; McMaster and Noble, 2005). While there is a significant amount of redundant storage, problems of sliver polygons will be minimal because each boundary is only taken from a single source. This is, therefore, a considerable enhancement on Kennedy *et al.*'s approach.

Rather than have snapshots of boundaries at key dates, another solution is to build a database containing a continuous record of boundary changes. Several countries have done this, using two distinct approaches: the *date-stamping* approach used by the Great Britain Historical GIS (GBHGIS) (Gregory *et al.*, 2002), and the *space-time composite* approach proposed as a theoretical structure by Langran (1992) and implemented by the Belgian Historical GIS (De Moor and Wiedemann, 2001).

The *date-stamping* approach handles time as an attribute in a similar manner to that described by Vrana (1989 and 1990). The problem is how to incorporate the need for spatial, let alone spatio-temporal, topology. Gregory *et al.* implement a date-stamping solution by storing all their spatial data in what they term *master coverages*. These are layers stored as ArcInfo coverages that take advantage of the coverage data model. In coverages, polygons are modelled using arcs to represent polygon boundaries and label points to represent the attributes of the polygons themselves. These can be stored in a coverage without topology, which can be added at a later stage. In the GBHGIS, label points represent the administrative units while arcs represent their boundaries. Boundary changes are handled in the manner shown in Fig. 6.4, which focuses on the registration district of Bromyard in Herefordshire. This was affected by three boundary changes, all of which resulted in the district losing territory. In one case, this was to Leominster; in the other two it was to Martley. The changes to Martley also affected the county boundary between Herefordshire and Worcestershire. Date stamps are stored as arc attributes (shown in Fig. 6.4d)

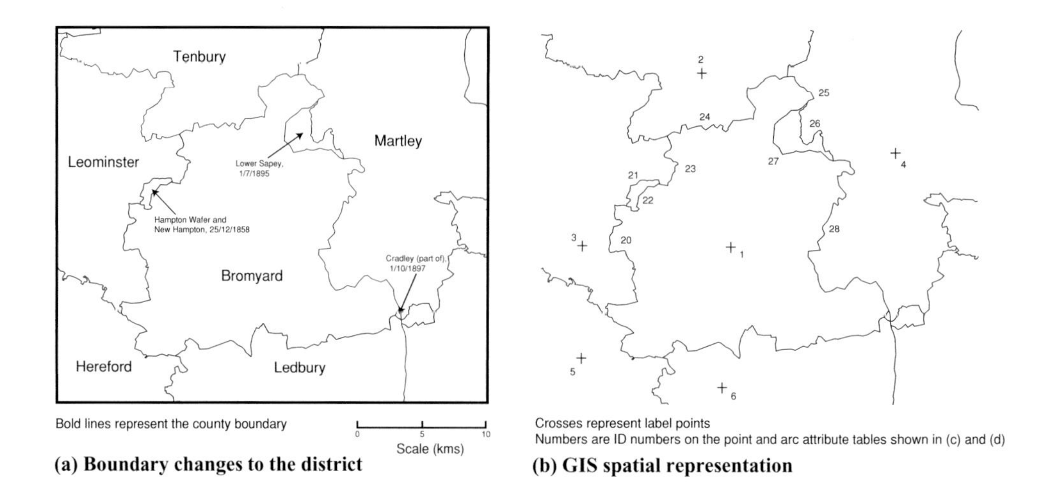

(a) Boundary changes to the district

(b) GIS spatial representation

ID	*Name*	*County*	*Start Date*	*End Date*
1	Bromyard	Herefordshire	0/0/0	0/0/5000
2	Tenbury	Worcestershire	0/0/0	0/0/5000
3	Leominster	Herefordshire	0/0/0	0/0/5000
4	Martley	Worcestershire	0/0/0	0/0/5000
5	Hereford	Herefordshire	0/0/0	0/0/5000
6	Ledbury	Herefordshire	0/0/0	0/0/5000

(c) Simplified label point attribute table

ID	*County 1*	*County 2*	*Un_cnty*	*Start Date*	*End Date*
20	Herefordshire			0/0/0	0/0/5000
21	Herefordshire			0/0/0	25/12/1858
22	Herefordshire			25/12/1858	0/0/5000
23	Herefordshire			0/0/0	0/0/5000
24	Herefordshire	W orcestershire	Y	0/0/0	0/0/5000
25	Herefordshire	W orcestershire	Y	0/0/0	0/0/5000
26	Herefordshire	W orcestershire	Y	0/0/0	1/7/1895
27	Herefordshire	W orcestershire	Y	1/7/1895	0/0/5000
28	Herefordshire	W orcestershire	Y	0/0/0	0/0/5000

(d) Simplified arc attribute table

Fig. 6.4 An example of the GBHGIS structure: the registration district of Bromyard. Note: all changes were transfers in which Bromyard lost territory. Start dates of 0/0/0 mean that the feature was in existence at the time the unit was created; end dates of 0/0/5000 mean that they were in existence when the unit was abolished.

that allow changes to boundaries to be handled. If a unit is created or abolished, this is handled by date stamping the label point with the date of the creation or abolition of the unit, with the accompanying boundary changes being encoded by date-stamping arcs. Therefore, if a user is interested in boundaries on 1 January 1859, only arcs and label points with a start date before this and an end date after it would be selected. These arcs and label points then have topology added, such that the resulting coverages contain polygons representing the administrative units on that date. The fact that place names are attached to the polygon attributes means that these can be joined to external data, such as census statistics, using a relational join as described in Chapter 2.

There are two problems with this approach. Firstly, it relies on a proprietary file format, namely ArcInfo's coverages, which are non-standard and may be phased out by ESRI, the company that owns the rights to this format. Secondly, there is neither spatial nor temporal topology in the master coverages, so there is no direct way of asking 'what area was adjacent to another?' at any date. The lack of topology also led to problems with quality control. When a layer is extracted for a particular date every polygon needs to consist of one label point completely surrounded by one of more arcs. Any errors in the date stamping of either arcs or label points is likely to mean that this is not the case, leading to errors in the extracted layers.

The Belgian Historical GIS described by De Moor and Wiedemann (2001) (see also VanHaute, 2005) implements a version of the space-time composite. At the core of their system is spatial data in polygon form that creates what they term a Least Common Geometry (LCG). Each polygon represents a fragment of an administrative unit. Reconstructing the administrative units for any date requires re-assembling the polygons, such that each fragment is allocated to the administrative unit that it belonged to on the date of interest. This is done using a series of relational database tables that provide information on which fragments belonged to which administrative units at all dates (Ott and Swiaczny, 2001). A simplified version of how this can be done is shown in Fig. 6.5, where there are two changes to administrative units: area *B* is transferred from 'Aplace' to 'Oldtown' in 1861 and then 'Oldtown' is renamed 'Newtown' in 1952. These are handled by two separate tables. The first table of attributes provides information on which administrative unit each polygon belonged to at every date. The second table gives the name of each administrative unit at every date. To reconstruct the administrative hierarchy on, for example, 1 December 1965, the first stage is to select the appropriate records from the LCG's attribute data. This tells us that polygon *A* belongs to administrative unit 1000 while *B* and *C* belong to 1001. This information can then be used to dissolve the boundaries of the LCG to produce the appropriate polygons for that date. The 'adm_codes' are then linked to the place-name table to tell us that in 1965 administrative unit 1000 was called 'Aplace' and 1001 was called 'Newtown'. Additional tables can be used to give the administrative hierarchy of higher-level units, and to link to attribute data from sources such as the census, or to qualitative data about the places at different dates stored on web pages. In this way, the structure is able to cope with changes in the shape of units, changes in their topology where new units are

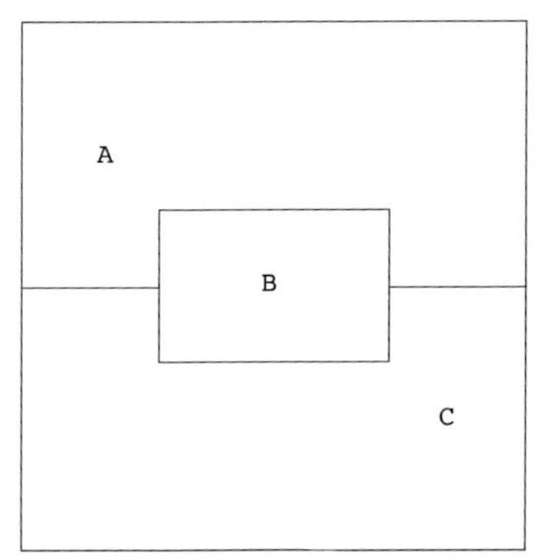

Spatial data

ID	*Start*	*End*	*Adm_code*
A	0/0/0	0/0/5000	1000
B	0/0/0	1/6/1861	1000
B	1/6/1861	0/0/5000	1001
C	0/0/0	0/0/5000	1001

LCG attribute table

Adm_code	*Start*	*End*	*Name*
1000	0/0/0	0/0/5000	Aplace
1001	0/0/0	1/1/1952	Oldtown
1001	1/1/1952	0/0/5000	Newtown

Place-name table

Fig. 6.5 Implementing a Least Common Geometry to create a temporal GIS. The spatial data consist of fragments of administrative units. The LCG attribute table is linked to this using the ID field. It states which administrative unit each fragment belonged to at every date. The place-name table gives the name of every administrative unit at every date. It is linked to the LCG attribute table using the field 'adm_code'. Start dates of 0/0/0 mean that the feature was in existence at the time the unit was created; end dates of 0/0/5000 mean that they were in existence when the unit was abolished. Thus, on 1 December 1965, 'Aplace' consists of polygon *A*, while 'Newtown' consists of polygons *B* and *C*.

created and/or old ones abolished, and changes in their attributes, including their names.

In many ways, the space-time composite structure used by the Belgian Historical GIS is more elegant than the date-stamping structure used by the GBHGIS. It does not rely on proprietary file formats to the same extent, as it uses polygon-based spatial data linked to a relational database. It is also less prone to errors in the logical consistency, as polygon topology already exists, although there is clearly the potential for errors in the attribute database, as, at every date, each polygon fragment must link to one, and only one, administrative unit code, and each administrative unit must have only one row of attribute data – such as place names – for each date.

Whether a date-stamping or space-time composite approach is used, the resulting database goes far beyond a representation of the original source or sources. Date-stamping and space-time composites allow for the creation of products that hold all three components of the data, taken from different sources. Spatial data are taken from maps, and temporal data may be taken from a variety of different sources, such as boundary change reports. These can be linked to a wide variety of both quantitative and qualitative attribute data, thus creating an integrated spatio-temporal database. No other technology, including paper, allows this to be done, and it has real potential to increase our understanding of spatio-temporal change using aggregate sources such as the census. For more information on how these have been used to advance knowledge, see Chapter 9.

6.4.2 *Changes to stand-alone features*

Many other situations that a historian may want to handle can be dealt with using a date-stamping approach where time is treated as an attribute. Healey and Stamp (2000) do this in their study of regional economic growth in Pennsylvania. For firms, represented as points, and railroads, represented as lines, the dates of their founding and closure are attached to the spatial features, as part of the attribute data. In this way, the development of the transport network and industrial development can be examined over time, and the links between the two can be studied.

The simplest way to implement this is with a single row of attribute data attached to each spatial feature. Multiple rows can also be attached to a feature, with each row having a start and end date. This allows situations, such as the name and ownership of a firm also changing over time, to be handled. A simplified example of this is shown in Fig. 6.6, where points representing the locations of three firms are shown. The attribute data are usually in the form of annual output data, as is the case for firm 1, Smiths. Where a firm opens or closes, this can also be included in the attributes, as happens with firm 2, which closes on 16 May 1871. Changes in names or ownership can be handled in the same way, thus the firm at point 3 changes from being 'Frasers'

Spatial data

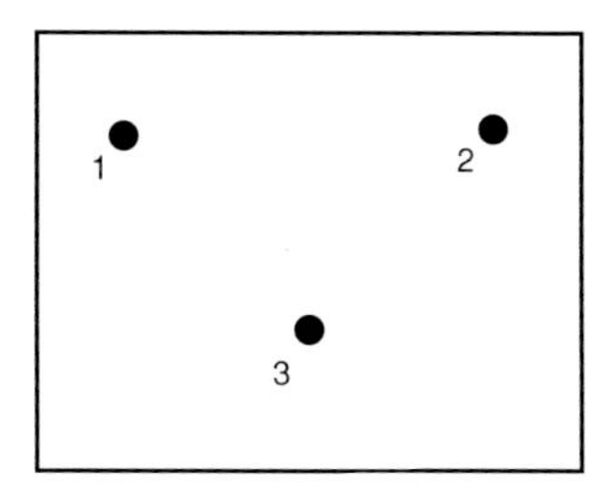

Attribute data

ID	*Start date*	*End date*	*Name*	*Output*
1	1/1/1870	31/12/1870	Smiths	870
1	1/1/1871	31/12/1871	Smiths	930
1	1/1/1872	31/12/1872	Smiths	990
2	1/1/1870	31/12/1870	Jones	405
2	1/1/1871	16/5/1871	Jones	115
3	1/1/1870	31/12/1870	Frasers	610
3	1/1/1871	31/12/1871	Frasers	540
3	1/1/1872	30/6/1872	Bloggs	205
3	1/7/1872	31/12/1872	Bloggs	365

Fig. 6.6 Time as an attribute with point data. Note that a more complex normalised structure could be used to reduce redundancy and increase flexibility.

to 'Bloggs' at the end of 1871. Handling time in this way allows spatial features to be created and abolished, and their attributes to change over time. This structure allows queries about what existed at a particular date, or how a location changed over time. It does not, however, allow objects that move over time to be tracked easily. Changes to stand-alone polygons can be done in much the same way, with date stamps being used to replace one polygon with another. Lines can also be handled in this manner; however, if a network is to be created, this complicates the situation, as topology is, again, required. To date, no-one has described doing this for the development of a transport network.

6.4.3 *An example of exploring change over space and time*

Diamond and Bodenhamer (2001) perform an analysis that uses GIS to explore change over space and time. Their chosen topic is 'white flight' in Indianapolis, Indiana in the 1950s, looking in particular at the extent to which Mainline Protestantism was involved in this process. They use three sources of data: 1950 and 1960 census data published at tract level, and data on churches, such as the size of the congregation, represented using points.

They wanted to test the widely held view that, as whites moved out of the racially mixed inner cities in the 1950s, Mainline Protestant churches also relocated, abandoning the inner cities both physically and spiritually. At the time, national commentators argued that there was widespread relocation of Protestant churches out of the inner cities, and that this showed that Mainline Protestantism was only interested in middle- and upper-class white congregations, rather than inner city African-Americans.

Simply mapping the changing racial mix in Indianapolis between 1950 and 1960 using census data confirms the commonly held belief of white abandonment of the inner city through the period. African-Americans increasingly dominate the inner city, while whites become more prevalent in the suburbs. Looking at changes in church locations over the period does not, however, lend much evidence to support the idea that Mainline Protestant churches moved out of the inner cities to get away from African-American neighbourhoods. By comparing the addresses of churches between 1951 and 1960, they found that only seventeen had relocated and, of these, only seven moved from the inner city to the suburbs. This only represented about 4 per cent of Mainline Protestant churches. Using overlay operations, they were able to examine the ethnic composition of the tracts that churches moved from. The seven churches fell in five tracts, as shown in Fig. 6.7. Over the period, the African-American population of two of these tracts expanded rapidly, a third grew modestly, and the

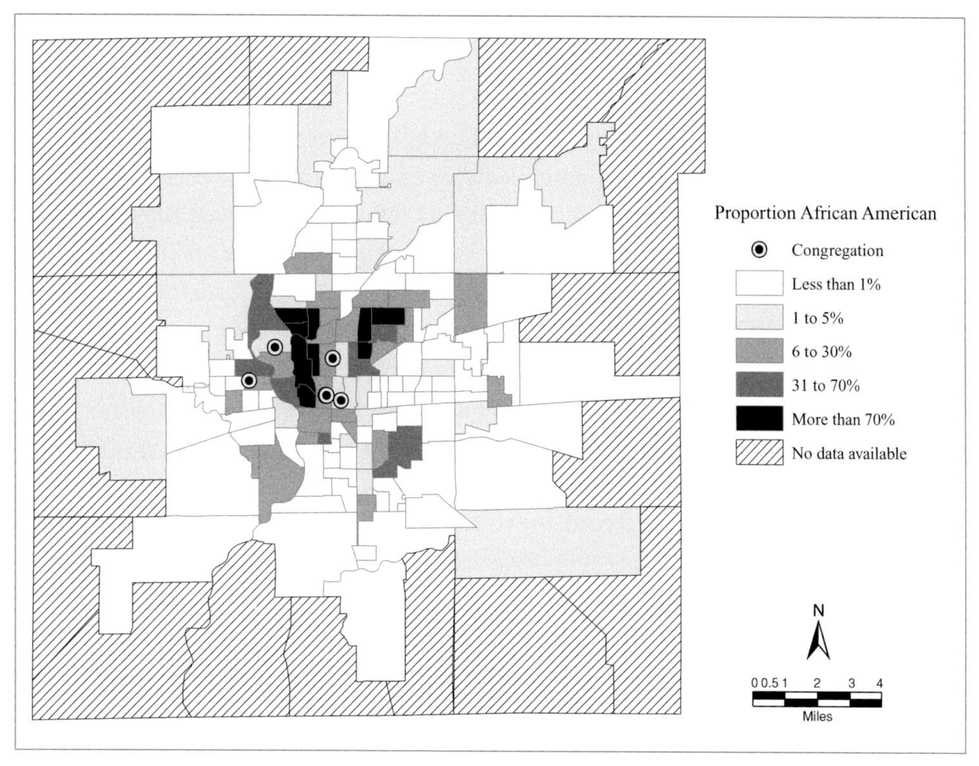

Fig. 6.7 The locations of Mainline Protestant churches that relocated in Indianapolis in the 1950s together with the African-American composition of the tracts that they left in 1950. Source: Diamond and Bodenhamer, 2001: p. 37.

other two only grew slightly. This suggests that race may not have been a significant motivating factor in the churches moving. As counter-evidence, however, the authors note that the tracts that the churches moved to all had negligible African-American populations.

The discussion above shows that, although time is not explicitly handled well by GIS data models and software, solutions can be developed to allow complex spatio-temporal databases to be created. These databases have the potential to integrate data from a wide variety of sources at different dates. Although not fully delivered on yet, this has the potential to deliver significant new knowledge on change over time and space. In this way, researchers should be able to use all three components of the data, and to be able to implement some of the approaches called for by writers from Langton to Massey. There are still, however, significant challenges to overcome. Fully integrating data over time will be referred to in the next section. Taking this further to generate new knowledge through analysis and visualisation will be dealt with in later chapters.

6.5 STANDARDISING DATA OVER TIME: THE ROLE OF AREAL INTERPOLATION

As described above, GIS allows the creation of spatio-temporal databases that allow the changing boundaries of administrative units, and data published by them, to be stored in an integrated structure. The usefulness of these data remains limited because boundary changes to the administrative units mean that it is still not possible to directly compare the data over time. This means that the trade-off described in section 6.2 still exists: that to explore data with their full amount of spatial detail, we can only look at a single date; comparing data over time requires aggregation to a set of administrative units whose boundaries can be assumed to remain constant. Therefore, data can either be explored with spatial detail, or over long time series, but not both. GIS and techniques pioneered in the field of spatial analysis allow this dichotomy to be resolved, based on a technique called *areal interpolation*, 'the transfer of data from one set (source units) to a second set (target units) of overlapping, non-hierarchical areal units' (Langford *et al.*, 1991: p. 56). The aim is therefore, to define the administrative units from one date as the target units, and interpolate data from all of the other dates onto these units. Although this can be used in any situation where data are published using incompatible administrative units, it is ideal for temporal data where the differences between units at different dates are relatively small. It is important, however, to stress that areal interpolation involves taking known data from the source units and *estimating* its values for the target units. These estimates are prone to error and thus must never be used uncritically. As will be described below, techniques can be devised to minimise this error, and to highlight data values that are more (or less) likely to contain error. If these techniques are not used, interpolated data must be used with extreme caution.

Areal interpolation involves performing an overlay between the source and target units to give the *zones of intersection* between the two. The degree of intersection is then used to estimate how much of each source unit's population[1] should be allocated to each target unit. The easiest way of doing this simply involves assuming that the population of the source unit is evenly distributed across its area. This is known as areal weighting and for count data its formula is:

$$\hat{y}_t = \sum_s \left(\frac{A_{st}}{A_s} \times y_s \right) \tag{6.1}$$

[1] Population can mean any data published for the administrative unit, including total population or any subsets of the population, such as those found in the census or in vital registration data.

Where $\hat{y}^t$ is the estimated population of the target zone, y_s is the population of the source zone, A_s is the area of the source zone, and A_{st} is the area of the zone of intersection between the source and target zones. Models similar to this can also be devised to handle data in the form of proportions, such as percentages, and can be used to estimate populations where the target zone is assumed to have a uniform density (Goodchild and Lam, 1980). This is shown in diagrammatic form in Fig. 6.8, where the populations of two source zones, *1* and *2*, are to be used to estimate the populations of three target zones *A*, *B* and *C*. The first stage is to overlay the source and target zones to produce the zones of intersection *M*, *N*, *O* and *P*. These have the combined attributes of the source and target layers, and the area of each new polygon which is calculated as part of the overlay operation. The user then estimates the population of each zone of intersection by dividing the area of the polygon by the area of the source polygon to give the zone of intersection's area as a proportion of the source area. This is then multiplied by population of the source zone to give the estimated population of the zones of intersection (Est. Pop. on Fig. 6.8). Finally, the zones of intersection are aggregated to target zone level using the target zone identifiers (TID on Fig. 6.8), so the estimated populations of the three target zones are 35, 45 and 70 respectively.

The difficulty with this technique is that, when dealing with human data, the assumption of an even population distribution across an administrative unit is clearly unsatisfactory. A variety of suggestions have been made to loosen this assumption, by using additional data that provide clues as to the distribution of y across the source units. Goodchild *et al.* (1993) suggest the use of a larger set of zones that can be assumed to have approximately even population distributions, although this type of situation is rare. Flowerdew and Green (1994) suggest adding attribute data for the target zones, using a statistical technique known as the *EM algorithm*. This will be returned to below. Most solutions, however, are *dasymetric* in their approach, in that they use a more spatially detailed dataset than either the source or target zones, to provide clues about the distribution of the population. Langford *et al.* (1991) use satellite imagery that is classified into types such as water, rural and urban. Norman *et al.* (2003) use the distribution of unit postcodes on the assumption that dense postcode distribution indicates a dense population. Reibel and Bufalino (2005) use the density of the street network on the same assumption.

The authors of all of these techniques claim that the approach they used improves the accuracy of their interpolated data. Very few, however, tackle the issue of error, beyond saying that their results are more accurate than simple areal weighting. It has also been shown that error will be reduced where source zones are relatively small,

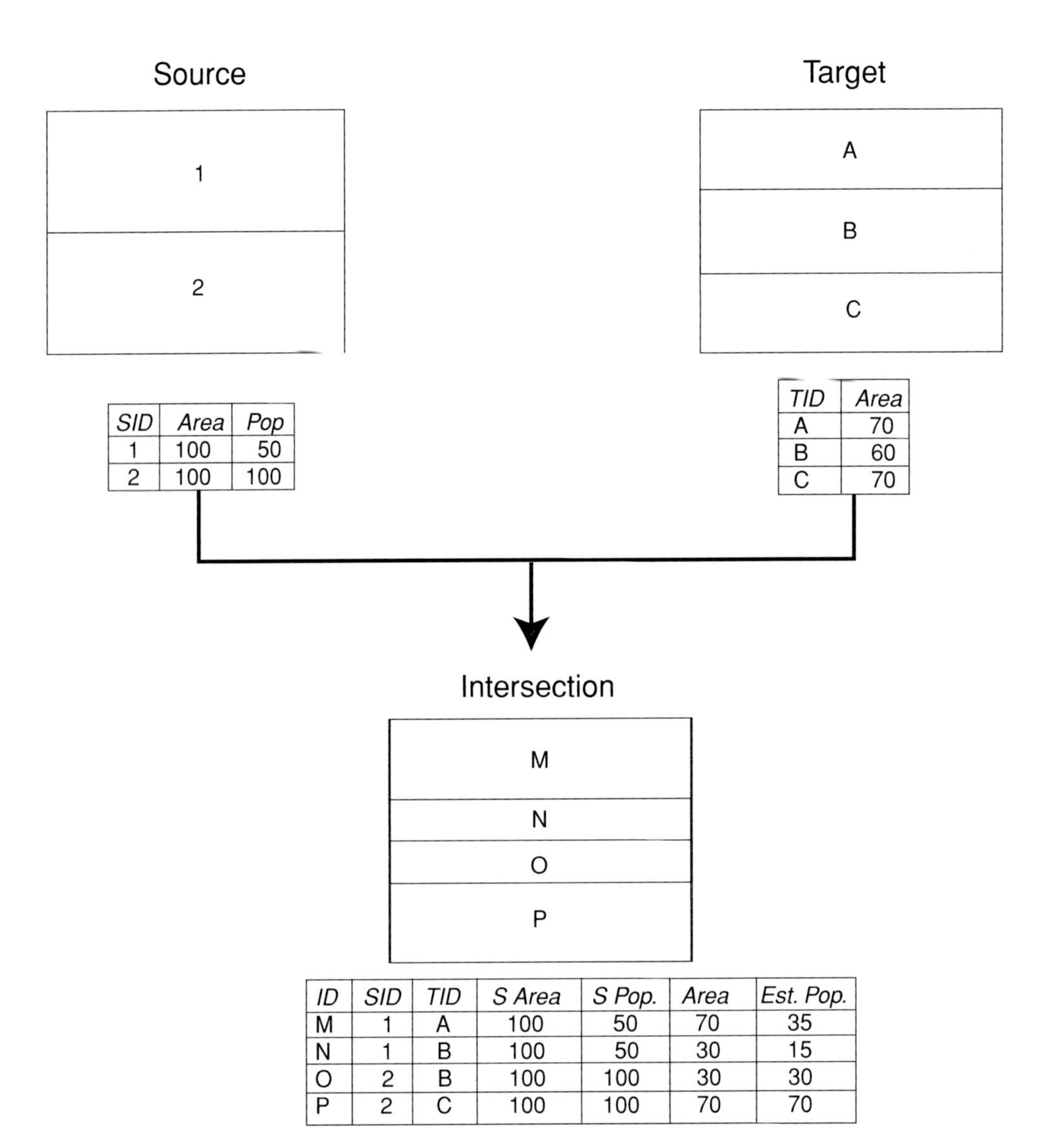

SID	Area	Pop
1	100	50
2	100	100

TID	Area
A	70
B	60
C	70

ID	SID	TID	S Area	S Pop.	Area	Est. Pop.
M	1	A	100	50	70	35
N	1	B	100	50	30	15
O	2	B	100	100	30	30
P	2	C	100	100	70	70

Fig. 6.8 Areal interpolation: estimating populations for overlapping polygons. The source and target layers are overlaid to calculate the zones of intersection between them. Attributes of the intersection layer include the IDs from the source and target layers (SID and TID respectively), area and population from the source layer (S Area and S Pop.), and the area of the polygons representing the zone of intersection as calculated as part of the overlay operation (Area). The final column (Est. Pop.) is the estimated population of each zone of intersection calculated as Area/S Area × S Pop. The final stage is to aggregate the values of Est. Pop. to target unit level, thus target $A = 35$, $B = 15 + 30 = 45$, and $C = 70$.

compared to the target zones, and that the relative shapes of the source and target zones will also affect accuracy (Cockings *et al.*, 1997; Sadahiro, 2000).

Gregory (2002b) explores the impact of interpolation error using a variety of approaches to interpolating historical data for England and Wales. Rather than using actual data, he uses 1991 census data aggregated to represent historical districts at different dates. He takes a number of variables, including total population, population without a car – which has an urban bias – and population working in agriculture. These are interpolated using techniques including areal weighting, a dasymetric technique that uses the total population of parishes to improve the accuracy of interpolation of other data only published at district level,[2] and an approach similar to Flowerdew and Green's, whereby data from the target unit were used to provide additional information. He shows that error varies widely between techniques and variables. Although the dasymetric approach improves on areal weighting, it can still lead to errors of up to 280%, nearly quadruple the actual value, while the EM algorithm-based approach reduces error to, at best, between 15 and 35%. Gregory and Ell (2005a) build on this to describe techniques to allow the user to select a technique likely to minimise the error. Gregory and Ell (2006) take this a stage further, to identify probable areal interpolation errors by exploring time series of interpolated data for unexpected changes. This opens the possibility of being able to use interpolated data in analytic research with the confidence that the impact of possible error can be identified and handled.

Areal interpolation is therefore a way of solving the problem of boundary changes, an issue that has seriously undermined our ability to make full use of sources, such as the census, to explore change over time. It is not, however, a perfect solution, as all interpolation techniques will inevitably introduce error to the data that they produce. The amount of error will vary according to the nature of the source data, the exact technique used, and the relative shapes and sizes of the source and target units. To make best use of interpolated data, the impact of this error must be considered and preferably must be explicitly addressed.

Gregory (2000) demonstrates the potential for using areal interpolation to allow traditional problems to be re-examined. As described above, he argues that long-term area-based studies of migration have been hampered by changing administrative boundaries. These have meant that researchers either used highly aggregated data, or were constrained in their temporal extent. For example, Friedlander and Roshier (1965) and Saville (1957) were able to produce county-based studies for the century from 1851 to 1951, but Lawton (1968), who used registration districts to give more

[2] This is the approach used by the Vision of Britain through Time project (Vision of Britain, 2005).

spatial detail, could only cover the period from 1851 to 1911. Gregory presents a case study looking at net migration at registration district-level from 1881 to 1931, a period that bridges a major re-organisation of administrative boundaries after 1911.

To conduct the analysis, he interpolated age- and sex-specific population data from the censuses of 1881 to 1931 onto a single set of registration districts. The 'basic demographic equation' (Woods, 1979) links population change, natural increase and net migration. For a given administrative unit, net migration, *NM*, can be calculated over a decade as:

$$NM = (p_{t+10} - p_t) - (B - D) \tag{6.2}$$

where p_t is the population at the start of the decade, p_{t+10} is the population at the end of the decade, and B and D are the number of births and deaths respectively over the decade. In England and Wales, there are sufficient data available to allow this to be used to calculate net migration for men and women in ten-year cohorts, using population data from the census and mortality data[3] from the *Registrar General's Decennial Supplements*, both of which subdivide data by age and sex. For example, if the census provides the number of women aged 15 to 24 at the end of the decade, and the number of women aged 5 to 14 at the start of the decade, and the *Registrar General's Decennial Supplements* provide the number of women in this cohort who have died over the decade, net migration can be calculated. Without interpolation, any population change caused by boundary changes will appear to be net migration. As all of the data have been interpolated, however, he is able to calculate net migration rates for a consistent set of spatially detailed units over the long term. The resulting data combine spatial and attribute detail, as they are at registration district level, and subdivide migrants by age and sex.

This analysis shows how areal interpolation can be used to create a major new dataset that can be used to provide further analyses of migration trends through the nineteenth and twentieth century. Doing this illustrates that the young migrated more than the old, and that migration patterns were different between men and women. He is also able to show that different areas within a county have markedly different patterns of net migration. While this is not surprising, in the past it has not been possible to quantify these differences.

Gregory *et al.* (2001b) follow a similar approach in analysing 100 years of poverty in England and Wales from the late-nineteenth to the late-twentieth centuries. They take three key quantitative indicators of poverty – infant mortality, overcrowded housing and unskilled workers – for key dates through the twentieth century. They

[3] Births are not required, unless we are calculating the very youngest cohort.

interpolate all of these data onto a single administrative geography, to allow them to be compared over time. The maps produced allow spatial patterns to be explored. The authors also perform a simple quantitative analysis that compares the mean rate of each indicator at each date, and what they term an 'inequality ratio' that compares the values for people living in the worst areas with those of people living in the best. The key point is that, without first standardising the data through interpolation, both of these measures would be meaningless, as they would be heavily influenced by the different administrative geographies used. The results of the study suggest that, while the mean rates of the indicators of poverty have declined sharply over the past 100 years, the inequality between the best and worst areas became significantly worse. This trend had been most marked in the last quarter of the twentieth century.

Both of these examples use areal interpolation to create new datasets that are then available for subsequent analysis. By manipulating the spatial component of the data, new datasets have been created that are explored, and have the potential for significant further analysis (see, for example, Congdon *et al.*, 2001). To date, the methodological issues in creating these datasets have largely been solved. Two issues remain: firstly, there is a lack of statistical approaches that can cope with analysing complex datasets over space and time simultaneously, and secondly, there is a need to apply these techniques fully, to advance our understanding of long-term change in society.

6.6 CONCLUSIONS

Time is not well handled by GIS databases, as the core data model used by GIS does not suit the addition of this extra component. As a result, in spite of over a decade of calls for time to be handled better within GIS, only limited progress has been made in incorporating temporal functionality into commercially available GIS software. While this may be considered a problem, it also opens up new opportunities. Historians tend to be suspicious of solutions to problems developed in other disciplines, and they are right to be so. Many of the issues associated with time in GIS are problems related to updating databases in response to changes over time. These are not typically relevant to historical GIS databases, which are static in nature, as they exist in historical sources that do not evolve. Therefore, although time is obviously a highly important concept to historians, many of the major difficulties in incorporating time into GIS databases, such as updating, scheduling and many of the quality control issues, are not relevant to them.

Within historical GIS, there has been significant progress in developing spatio-temporal databases. Particularly, progress has been made in the development of national historical GISs. The development of these systems shows that, with ingenuity, solutions can be found that allow change over time to be effectively incorporated into a complex GIS database. The use of areal interpolation shows that, once created, these systems can be used further to manipulate data, to allow it to be directly compared over time – something that has not previously been possible.

For many years, historical geographers have been calling for a better understanding of change over time and space; however, the complexity of handling data through all three components simultaneously has seriously limited their ability to do this. Developments within historical GIS have advanced the agenda to a position where it is now possible to develop databases that allow all three components of the data to be effectively represented. This is a major step forward, and has significant implications for our understanding of change within society. The implications of this go beyond the discipline of history, and move into contemporary geography, as it allows questions such as 'how did we arrive at the situation we are in at the present?' to be answered. It must be stressed, however, that although techniques have been developed to create databases that hold thematic, spatial and temporal information, this is only the first stage in the process. Manipulating them to make them directly comparable through, for example, areal interpolation and, potentially, handling the error that this introduces is a second stage. Even after these two stages, however, all that has been developed is a database. Turning a database into knowledge is a further challenge. The reason data tend to be simplified in the way described by Langran and Chrisman (1988), whereby only one of the three components is explored in detail, is, in part, down to the human mind's difficulties in handling space and time simultaneously. Exploring data through space and time to develop an understanding of both remains complex. Further work is still required in developing visualisation (Chapter 5) and analysis (Chapter 8) techniques that will allow this to happen. GIS has, however, moved us a significant step closer to being able to understand the complexities of change over time and space.

7

Geographic Information Retrieval: historical geographical information on the internet and in digital libraries

7.1 INTRODUCTION

So far, this book has been concerned with the overlap between two very different sub-disciplines: historical geography and GIS. In this chapter, a third subject is used extensively: library science. Large digital libraries and the internet have the potential to make data available to historians in volumes that are still difficult to imagine. A major problem that this will inevitably lead to is simply finding relevant data. If a user is interested in a particular place, then clearly GIS has a role to play in helping to find information about that place. Library science has experience in designing catalogues and, more recently, search engines that allow users to find the resources they require. Adding searching by place requires the use of locational information, either in the form of co-ordinates, or as place names, or both. This requires an overlap between GIS and library science that has led to a growing inter-disciplinary field, sometimes known as Geographical Information Retrieval (GIR) (Larson, 2003). This is concerned with how best to use the locational information within datasets to find data held in digital libraries or across the internet.

GIR is strongly linked to two emerging fields: grid technologies and e-science. Grid technologies are based on the potential offered by the high-bandwidth connections now available over the internet. There are three implementations of the grid. The *data grid* makes use of computer networking to link large digital resources from around the world quickly and efficiently. Scholars can use the data grid to access, for example, large image files, such as areal photographs, high quality scans of historical maps, or large statistical databases. This has been made possible by the rapid increase in the speed of internet connections, which makes large internet resources almost as readily available as they would be if they were stored on a local network. The *computation grid* provides access to high-specification computing power and software routines in a distributed environment, again using the speed of the modern internet. This means that a researcher does not need to have a particular piece of software or a particularly high-powered computer available to them, as they can access the required resources online. The *access grid* provides virtual research environments, allowing scholars to collaborate over the internet. E-science provides tools and routines to facilitate the efficient use of the grid (Berman, 2003). Both groups of technologies are well established in the sciences, but have yet to make a significant impact in the humanities. The data grid, with its ability to link disparate and distant data resources, has particular potential.

This chapter explores the use of GIS over the internet, with a particular emphasis on finding and retrieving information. Section 7.2 of the chapter explores how a GIS database can be put on the internet so that it can be disseminated to a global audience. This approach effectively puts a subset of the tools offered by standard GIS software into a web browser, and allows users to explore the dataset in a conventional GIS manner, using mapping of layers and spatial and attribute querying. The potential for Geographical Information on the internet goes far beyond this. The biggest challenge posed by the internet is the problem of finding relevant data. Searching by theme is one approach, but users may also want to search by location, or by date. Searching by location involves handling Geographical Information effectively but is not necessarily simple. Section 7.3, therefore, takes the approach that spatial data are a form of metadata that allow the user to find the data about the place that they are interested in. This can occur at two levels. A single dataset can be searched for all of the individual records that are at, near, or intersect with a certain location, alternatively, a catalogue of datasets can be explored to find all of the datasets at, near or intersecting with a certain location.

A limitation to this is that it requires the search location to be defined using co-ordinates. Many datasets are organised by place name, rather than by co-ordinate. Gazetteers, otherwise known as *thesauri*, may be created to convert from a place name to a co-ordinate-based location. The Getty Institute's Thesaurus of Geographical Names is a good example of an extensive gazetteer that can convert between place names and the co-ordinates of a centroid representing their locations (Harpring, 1997). There are, however, many challenges for gazetteers, particularly with historical data: place names change over time and between sources; the area referred to by a place name may change; places are often hierarchical, especially where administrative or political units are used; and a point may not be an effective way of representing a place. Section 7.4 looks at gazetteers, and explores how they offer an alternative database-driven approach to handling locational information that complements the GIS approach described in section 7.3.

Although their uses are not restricted to resource discovery, the combined ability of spatial data to act as metadata, and gazetteers to convert between co-ordinates and place names, opens up a new potential for finding historical data on the internet. This gives the historian unrivalled access to information about the place, time and topic that they are interested in. The implications of this for the discipline are still very much in their early stages, but we argue that they may be among the most significant that GIS has to offer to historical research.

7.2 BASIC HISTORICAL GIS ON THE INTERNET

In its simplest form, an *internet GIS* provides a user with basic GIS functionality through a web page. The user opens the web page in the usual way, and a plug-in or applet will appear that provides basic GIS browsing functionality, such as the ability to pan around a map, to zoom in and out, to turn layers on and off, and perform attribute and, perhaps, spatial queries. This allows a person who has created a GIS database effectively to publish it on the internet and give a global audience access to it (Harder, 1998; Peng and Tsou, 2003). A variety of software are available to do this. ESRI's Arc Internet Map Server (ArcIMS) (ESRI, 2005) is a leading commercial product from a GIS software developer, the Oracle Spatial module of the Oracle relational database management software provides a more database-oriented approach (Oracle, 2005), and certain free products, such as TimeMap, are also available (TimeMap, 2005).

A difficulty with internet GIS is that, as with all internet applications, responses to an internet query must be close to instantaneous, otherwise the user will lose interest. This poses particular problems with GIS data, as spatial data volumes are often large, and many GIS data-processing operations that a user may wish to undertake, such as overlay and buffering, are computationally complex, so processing times may be unacceptably slow. For this reason, the functionality offered by internet GIS software is often limited, and people wishing to publish GIS data on the internet must think carefully about how to minimise response times to queries. This may require significant technical expertise if complex data are used, particularly if these are polygons.

A good example of a simple GIS on the internet is provided by Schaefer's (2004) online GIS of the Tithe Survey of England and Wales. The tithe was a survey of the country taken for taxation purposes that focused on land ownership and agricultural wealth. It followed the Act of Commutation of the Tithes in 1836. It is a source that has been extensively used by historians, as it provides a comprehensive survey of the country in the early stages of the agricultural and industrial revolutions (Kain and Prince, 2000). The structure of the survey is well suited to GIS, as it combines maps of fields and other land parcels, with a schedule of information about these areas. The schedule includes statistical data, such as areas, owners, tenants, and values, in tabular form. The maps are linked to the schedules using ID numbers that appear on both sources. The original tithe maps are stored in the National Archives (formerly the Public Record Office) and a variety of County Record offices. They are often in poor condition and can be up to 10 m^2, making them very difficult to handle. Schaefer and his colleagues have been capturing data from tithe maps and surveys, and turning them into a GIS, with the maps being stored in shapefiles as vector data, mainly polygons, and the schedule stored in a relational database management system. These are disseminated over the internet in ways that allow the user to pan and zoom, access schedule information about individual areas, and query data from the schedules. Maps can also be printed or saved as image files. The site is available through the University of Portsmouth's website.[1] To date, only a limited coverage of the country is available. An interesting feature of the system is that it allows data to be uploaded to the site from remote sites, such as record offices, to allow them to be disseminated directly. This has the potential to significantly speed up the dissemination process, although for quality control reasons this is not publicly available.

This is a good example of a simple online GIS. Like a conventional GIS, it has the advantage of combining map-based information with attribute information in tables to create an integrated and more usable product. Putting it on the internet allows inaccessible information, stored in a variety of record offices and archives, to be made available to a global audience through a single website, without the need for any specialist software. There are, however, some difficulties. One criticism is that the data on the website are removed from the source documents. This may be undesirable in itself, but it also raises questions about the accuracy of the GIS transcription. This could be handled by effective documentation and metadata (see Chapter 3), but this is not present on this site.

A second and contrasting example of an online GIS is provided by the Gough Map project at Queen's University, Belfast.[2] The Gough map is an early map of Great Britain held in the Bodleian Library, Oxford. It is believed to date from the fourteenth

[1] See www.port.ac.uk. Viewed 19 June 2007.
[2] See www.qub.ac.uk/urban_mapping/gough_map. Viewed 19 June 2007.

century. The map has been scanned and put on the internet, with enhanced features that allow the user to, for example, identify place names and their modern equivalents, measure distances between features, and turn additional layers on and off. These have been digitised from the scan to allow features such as roads, settlements and rivers to be easily identified and queried. This is thus, primarily, a raster system based on a single source. By placing it on the internet, the map is made available to a wide audience, and the additional features allow further study of what the map shows in ways that would not be possible on a delicate and irreplaceable original.

7.3 USING LOCATION AS METADATA

As has been stated earlier in the book, one of the key advantages of GIS is its ability to use location to structure and integrate data, particularly when the data are taken from disparate sources. Digital libraries in particular, and the internet more generally, store large volumes of data, much of which have a spatial reference. This presents two significant challenges that GIS can help to overcome: firstly, finding suitable data about a place, and secondly, integrating these data with other suitable data.

A key concept to understanding why GIS and resource discovery complement each other is the *spatial footprint*. At the core of Geographical Information is the use of co-ordinates to express where a feature is on the Earth's surface. Usually in GIS, this is done using spatial data that represent where a feature is, to a high degree of precision, if not accuracy (see Chapter 4 for the distinction), using points, lines, polygons, or pixels. In many of the applications described in this chapter, having an accurately defined location is neither possible – due to uncertainty or vagueness in the source, which is necessary, as the aim is to find data in approximately the right location – nor practical – it may lead to slow processing times. A spatial footprint is thus used as an approximate location for a feature. While polygons and lines can be used, it is more common to use a point or a *bounding box*. A bounding box is an area that can be defined using only two co-ordinates, the south-west and north-eastern corners of the area of interest. If, for example, we are interested in selecting features in a cultural region such as the south-western US, one approach would be to define which states are to be included in our definition of the south-western US, and attempt to select all of the features that lie within the polygon that this creates, using an overlay operation (see Chapter 4). This is impractical, as extracting the polygons that define the south-western US, and using an overlay operation, are both slow processes, and it is also unlikely to give a satisfactory definition of the south-western US. Using a bounding box may well give an equally satisfactory definition of the place of interest,

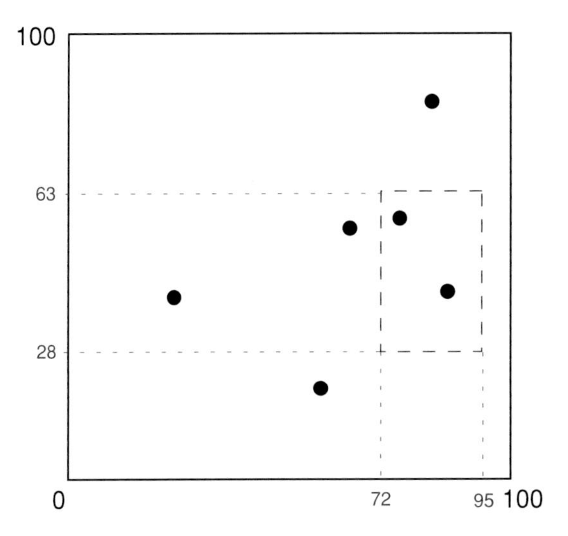

Fig. 7.1 Selecting features using a bounding box. As a box can be defined using only two co-ordinate pairs – in this case (72,28) and (95,63) – selecting features that lie within the box are easily processed as a database query.

and will almost certainly do so far more quickly. Firstly, it only requires the use of two co-ordinates to define south-western US, and secondly, extracting the features lying within this area can be done using a simple database query, rather than a complex overlay operation. Thus an attribute in a conventional database can be used instead of a spatial query on a GIS layer. An example of this is shown in Fig. 7.1. If we want to select the points that lie within the bounding box, which may be drawn on-screen by a user, processing this in a conventional database, as opposed to a GIS, is easy and efficient. The query simply becomes: 'select all the points where *x* is between 72 and 95 and *y* is between 28 and 63'. A GIS allows these co-ordinates to be defined by drawing a bounding box on-screen.

This opens up the potential of using an internet GIS browser as a tool to search a catalogue that works in tandem with more conventional searches, such as keywords. A conceptually simple but highly effective example of this use of spatial data as metadata to search for individual records within a historical database is the International Dunhuang Project (IDP) at the British Library (IDP, 2005). This is a digital library of over 100,000 manuscripts, documents and document fragments, and other artefacts dating from the fifth to the eleventh centuries. These were found in Dunhuang in north-west China and at other sites along the Silk Road. They were excavated in the early twentieth century by archaeologists from several countries, including Britain. This resulted in the contents of the sites being dispersed among institutions in China,

Britain, France, Japan and Russia. The artefacts, many in poor condition, cannot be made generally available to the public, so the IDP has created an integrated digital library of them. In addition to the manuscripts, the project is also digitising and geo-referencing the notes, pictures and expedition maps taken by Sir Aurel Stein, a British archaeologist responsible for much of the excavation. At its core, therefore, the IDP takes a large and geographically dispersed repository of historical artefacts stored at different locations, and creates a database of images and scans that is made publicly available through a single catalogue. This even permits fragments of the same manuscript excavated from separate sites by different teams, and stored in dispersed locations, to be digitally reunited.

The IDP catalogue includes mapping functionality that allows users to search for manuscripts using location, by allowing spatial data to be used as metadata. The individual items in the database have been geo-referenced as points, and this aspect of the catalogue is presented to users through an internet GIS browser with background maps providing contextual information. Users can, for example, select all of the artefacts in a certain area, or select all of the artefacts found near another artefact. This, in turn, stresses the geographical nature of the resources, something that is usually removed by conventional catalogues.

Another example of this approach is given by the Sydney TimeMap project (Wilson, 2001; see also Chapter 5). This is based at the Museum of Sydney, a museum that traces the development of the city from the time before European settlement in Australia to the present. The Sydney TimeMap is a museum display that attempts to put many of the artefacts held in the museum back into their spatial and temporal context. At its core is a series of maps of Sydney at different stages of its development. Added to this are over 500 sketches, images, engravings, etchings and photographs, all of which have been given a point location and a date. The TimeMap software (TimeMap, 2005) allows the user to select features by time and location and thus are able to explore where features were in relation to each other, in relation to the maps of Sydney, and how they evolved with Sydney over time. Wilson claims that the user response to this map-based approach to exploring the artefacts has been strongly positive, as it is a highly intuitive way of interacting with the data. Unfortunately, to encourage visitor numbers, the database is not publicly available over the internet, but is only available in a kiosk in the museum. Nevertheless, this gives an interesting idea of how spatial data can be used to create a catalogue that is both easy to use and highly effective in stressing the historical and geographical nature of the artefacts.

Other examples of digital libraries that use spatial data as metadata include: the Perseus Project (Smith *et al.*, 2000), a library of humanities resources, covering texts from Egypt in the third millennium BC to nineteenth-century London; the

Alexandria Digital Library (ADL), a map library that has evolved out of the University of California, Santa Barbara's Davidson Library (Goodchild, 2004; Hill and Janee, 2004); and the Rumsey collection, a private collection of historical maps (Rumsey and Williams, 2002; see also Chapter 2).

As well as finding individual records within a dataset, Geographical Information can also be used to find entire datasets using spatial information held in the dataset's metadata. The Dublin Core Metadata Standard (Dublin Core Metadata Initiative, 2005; see also Chapter 3) is not well equipped to handle information in this form; however, the Federal Geographic Data Committee's (FGDC) Content Standard for Digital Geospatial Metadata includes an element termed the *spatial domain*, which allows the geographical extent of the dataset to be described. This can be done using either a point (a single co-ordinate pair), a bounding box (two co-ordinate pairs that provide the south-western and north-eastern limits of the dataset), or one or more polygons (FGDC, 2005 – this is described under standard FGDC-STD-001–1998). Another metadata standard of relevance to historical GIS is the Data Documentation Initiative (DDI) (DDI, 2006). This was developed specifically to describe social science datasets (Block and Thomas, 2003). It is not explicitly designed to handle Geographical Information, but has been used by both the US National Historical GIS (see Chapter 6) and the Vision of Britain project (see Chapter 5). These implementations are described by Fitch and Ruggles (2003) and Southall (2006) respectively.

Metadata for many datasets can be held in a metadata catalogue, which is, effectively, a database that contains information about datasets. The metadata catalogue is usually accessed over the internet through a *portal*, a website that contains the metadata, but not necessarily the data that they refer to. Users access the portal and use the metadata to find data that may be relevant to them. There are then links to the site where the data are stored. It may be possible to access data directly through these links, or there may be access constraints to protect the data owner's intellectual property rights. Portals may, potentially, allow data owners to upload their metadata to the portal to allow others to access it, or may limit their catalogues only to items that they are interested in. The portal is thus an access point to a distributed data library. It provides access to datasets stored in a variety of places through a metadata catalogue which provides users with the appearance of a single site from which they are able to find data. The advantage of this architecture is that running a portal is cheaper and more flexible than storing all of the data centrally, as most of the storage and maintenance costs are borne by the data owners who have the additional advantage of retaining control over their own data. A disadvantage, as discussed in Chapter 3, is that, as the data are stored on remote machines, they may be moved, deleted, or become obsolete, and the long-term preservation costs remain with the owner. Therefore, while making data available through a portal is often a good dissemination strategy, it does nothing to ensure long-term preservation.

If the metadata have a spatial footprint, then metadata catalogues can be searched for datasets in much the same way as the digital library catalogues, described above, can be used to search for individual records. The Electronic Cultural Atlas Initiative's (ECAI) Metadata Clearinghouse is a good example of a portal for historical data that uses spatial metadata and map-based searching (Lancaster and Bodenhamer, 2002; Larson, 2003; ECAI, 2005). The Metadata Clearinghouse at the core of the portal is based on Dublin Core, with additional fields developed by ECAI that provide spatial metadata either through a point or a bounding box. Data owners can freely upload their metadata to the Clearinghouse. Some of the datasets themselves may also be stored in the Clearinghouse, while others are stored remotely. Once the metadata are in the Clearinghouse, users can search for datasets using a map-based interface that allows them to pan and zoom around a map, in addition to conventional catalogue search tools. It also has a time-bar that allows resources to be selected by date. In this way, a user can easily visually explore all the datasets that cover the area that he or she is interested in, and download them to their own computer for further analysis. The Geography Network (Geography Network, 2005) is another example of a spatial metadata portal that allows data to be located spatially, although this is not aimed explicitly at historical resources.

Increasingly, there are moves to make metadata catalogues *inter-operable*, such that if a query is entered into a portal, it searches a variety of metadata catalogues and returns all of the results. This has enormous potential for resources discovery over the internet, and is bound to increase significantly in the near future as uses of the data grid increase in the humanities.

7.4 GIS AND PLACE-NAME GAZETTEERS

The use of spatial data as metadata relies on being able to define a place using co-ordinates. Although this is the easiest way of approaching location from a computing perspective, humans are far more comfortable with place names. Place names are, however, problematic. The same name may refer to different entities in different contexts – for example, New York may refer to a city or a state, or Durham may refer to a city or a county. They may also be ambiguous – for example, does London refer to a precisely defined administrative area, or a more fluid cultural place? Different languages may refer to the same place using different names, such as

x	*y*	*Parish*	*District*	*County*	*Country*
453648	283942	Lutterworth	Lutterworth	Leicestershire	England
453924	282009	Cotesbach	Lutterworth	Leicestershire	England
448815	283942	Monks Kirby	Lutterworth	Leicestershire	England
449230	273586	Bilton	Rugby	Warwickshire	England
444673	284632	Withybrook	Foleshill	Warwickshire	England
440807	284770	Shilton	Foleshill	Warwickshire	England

Fig. 7.2 A simple gazetteer structure. This gives the centroids and names of parishes along with which district, county and country they are found in a single table.

Germany, Deutschland or Allemande. Even if names are in the same language, different historical sources may use different names or different spellings to refer to the same place. Many place names, particularly those associated with political or administrative units, will either change over time, or the area that they refer to will change over time. Thus Persia, India and Germany all refer to different areas at different dates. To attempt to resolve this, *place-name gazetteers* may be used. These are database tables that attempt to disambiguate place names, and associate them with a spatial footprint on the Earth's surface, usually in the form of a point, a bounding box or a polygon (Buckland and Lancaster, 2004). A gazetteer, therefore, is a way of translating place names to co-ordinates, which opens up the potential for a user to search spatial data by entering place names. A simple use of a gazetteer to standardise spellings of place names was given in Chapter 3 (see, in particular, Fig. 3.6). This is developed here further, with particular reference to how gazetteers may be used over the internet.

Gazetteers are often used to store information about hierarchies of places. This is particularly true of gazetteers of administrative units that have clearly defined nested hierarchies. For example, in Britain there has traditionally been an administrative hierarchy that goes from small administrative units such as parishes, through middle-sized ones such as districts, and up to large ones such as counties, and perhaps even the countries of England, Scotland and Wales. By incorporating this type of information into a gazetteer, it becomes possible to answer queries such as 'what parishes constituted the county of Warwickshire?' or 'which district does this parish lie in?' This type of information is easily stored in a conventional database table, as shown in Fig. 7.2. In this example, the spatial footprint used is simply a point. For each parish, we have the co-ordinate pair and the district, county and country that the parish lies in. Therefore, it tells us that Bilton is a parish in Rugby, which is in the county of Warwickshire in England, and that the parish is represented as being located at (449230,273586). It could also be used, for example, to give a list of all

x	y	Parish	District
453648	283942	Lutterworth	Lutterworth
453924	282009	Cotesbach	Lutterworth
448815	283942	Monks Kirby	Lutterworth
449230	273586	Bilton	Rugby
444673	284632	Withybrook	Foleshill
440807	284770	Shilton	Foleshill

District	County
Lutterworth	Leicestershire
Rugby	Warwickshire
Foleshill	Warwickshire

County	Country
Leicestershire	England
Warwickshire	England

Fig. 7.3 A normalised gazetteer structure. This structure contains all of the information held in Fig. 7.2, but reduces the storage capacity needed by 'normalising' the data. To find out which parishes lie in Warwickshire would require a relational join between the left-hand and central table on the district-name field.

Standard	Alternate
Monks Kirby	Monks Kirby
Monks Kirby	Monk's Kirby
Monks Kirby	Monks Kirkby

Fig. 7.4 A gazetteer table to standardise the spellings of place names. This table fragment could be added to the structure given in Fig. 7.3 to allow variations in the spellings of parish names to be handled.

the parishes in the district of Lutterworth. This, in turn, could be used to define a bounding box that would enclose all of the centroids within this district.

Although the structure in Fig. 7.2 contains all of the required information, it is inefficient, as, for example, the county in which each district lies is repeated many times, as is the country in which each county lies. Using a relational database allows this to be simplified, as shown in Fig. 7.3. Here, the gazetteer has been split into a number of tables: one that gives the spatial footprint and the name of the district that each parish lies in, a second that gives the name of the county that each district lies in, and a third that gives the name of the country that each county lies in. This greatly reduces the amount of storage required, by reducing the amount of redundant or duplicate information (Date, 1995; Hernandez, 1997). Therefore, we now only need to store the information that 'Warwickshire' is in 'England' a single time, rather than once for every parish in Warwickshire. At the same time, if we want to find out which parishes lie in Warwickshire, this can be achieved by joining the table containing country names to the table that holds the parish names, using a relational join on the district names.

Where alternative spellings of place names exist, these can be handled in the same structure by adding further tables to the design described in Fig. 7.3. These tables would hold the standardised names in one column, and the alternative spellings in a second. Figure 7.4 shows a possible fragment of this for parish names. Using relational joins, it would then become possible to answer queries such as 'which district does the parish of Monk's Kirby lie in?' This would be answered by first converting from

'Monk's Kirby' to 'Monks Kirby', the standard version of its name, using the table shown in Fig. 7.4, and then querying the gazetteer structure shown in Fig. 7.3 to find out that it is in Lutterworth. In this manner, complex information about places, their hierarchies and spellings, can be built up using conventional database technology. For internet applications, this has the advantage of using standard forms of software that offer rapid processing speeds, rather than GIS, which is still specialised and slow.

In the examples given above, the spatial footprint is restricted to a point. It can easily be extended to be a bounding box by adding a second co-ordinate pair to give the south-west and north-east limits of the administrative unit. If a polygon is required, this becomes more complicated, as most conventional relational databases do not handle polygon boundaries well. The best solution to this may be to link from a relational table of place names to a GIS layer that gives the place names as attributes, and their polygons as spatial data. This means that we are now using the best software for each of the required purposes. It has the disadvantage of significantly adding to the complexity of the database architecture. It will also significantly add to the computer processing time required to handle a query. Relational databases with spatial functionality such as Oracle Spatial (Oracle, 2005) provide another possible solution.

A further complexity for gazetteers is change over time, which may affect whether the place exists or not, where the place is located, how its name is spelt, and where it lies in an administrative hierarchy. Many of the issues of handling time in GIS were dealt with in Chapter 6. Here, it is sufficient to note that time can be handled by adding start- and end-date fields to the gazetteer. This, again, adds to the complexity of the structure, but also adds much to its utility. A further complexity is that, much like places, periods of time may also be ambiguous, and their validity may vary from place to place. For example, 'the industrial revolution' refers to different periods of time in different countries. Again, gazetteers can be devised to handle these, but we are some way from a comprehensive system that allows this.

An example of a system that uses GIS and gazetteers together that is currently available is the Vision of Britain through Time project (Southall, 2006). A user using the site is first prompted to type in a postcode. This is converted into a district based on the co-ordinate of the postcode's centroid. A variety of data can then be returned about this district, including which historical districts it overlapped with and the related census data. Historical maps centred on the postcode can be drawn, and information about nearby places retrieved from John Bartholomew's 1887 gazetteer of the British Isles and a number of travellers' tales. Integrating data in this way required the combined use of GIS, gazetteers and spatial metadata.

In theory, the technology exists for queries such as 'find me all information on villages that exist within ten miles of the River Severn' or 'which nineteenth-century Poor Law unions overlapped with the modern county of the West Midlands[3]?' to be entered into a portal, and a wide range of data from across the internet to be returned. In reality, we are still some way from this, although research projects exist that are moving towards this (see, for example, Hill, 2004; Larson, 2003; Reid *et al.*, 2004). To aid the user further, being able to also search by theme and by date, and have the results ordered by how well they appear to match the search criteria, is also desirable. Jones *et al.* (2003) offer a solution to this.

7.5 CONCLUSIONS

The internet is still evolving rapidly, GIS on the internet lags behind this, and historical GIS on the internet is still very much in its infancy. Nevertheless, there are already projects that demonstrate the potential impact of the internet and GIS on historical scholarship. Schaefer's (2004) work on the Tithe Survey shows the potential for disseminating GIS data over the internet. The International Dunhuang project (IDP, 2005) and the Sydney TimeMap project (Wilson, 2001) show what can be achieved to integrate disparate physical objects in a single catalogue, and, in the case of the IDP, make these data available to a global audience over the internet. The use of spatial data as metadata, combined with a mapping front-end to allow users to query the database, enables them to explore it in a manner that stresses the underlying geography inherent in the data. Conventional approaches to cataloguing this data would render this impossible. The Vision of Britain through Time project (Southall, 2006) goes further than this, making use not only of spatial metadata, but also of gazetteers, to allow data from very different sources, including maps, statistical information and textual descriptions of places, to be integrated and queried.

All of these projects effectively place a GIS database on the internet, but are still only single sites that only contain the information that the database creators have decided to include. The full potential of historical GIS on the internet is the potential for researchers to integrate data from multiple disparate sources, and either to explore them over the internet, or download them to their computer for further analysis. Doing this requires co-ordinate-based spatial metadata to allow information about

[3] The county of the West Midlands was formed in 1974 from parts of the counties of Warwickshire and Staffordshire. It did not exist before that. Poor Law unions were a type of administrative unit used to administer the New Poor Law. They ceased to exist in the early twentieth century, thus at no point did Poor Law unions and the county of the West Midlands exist at the same time.

places to be discovered and to be integrated with other data that also contains co-ordinates. The ECAI Metadata Clearinghouse (Lancaster and Bodenhamer, 2002; ECAI, 2005; Larson, 2003) goes some way towards this. It provides a metadata catalogue that includes data from a number of disparate sites that can be integrated through the Clearinghouse using the spatial metadata. These can then be downloaded to the individual researcher's computer for further analysis. The Clearinghouse, therefore, provides a tool that allows researchers to integrate data themselves, rather than using pre-prepared data of the type offered by the IDP or the Vision of Britain projects.

The potential for historical GIS over the internet is still, however, largely unrealised. There are two areas where there is significant further potential. Firstly, at present, there are two distinct levels of spatial metadata: *dataset-level* metadata that describes the spatial extent of the entire dataset, and *record-level* metadata that describes the spatial extent of each record within a dataset. In an ideal world, this dichotomy would not exist or would at least be reduced, such that, for example, a local historian could find information from national datasets that are of relevance to his or her local area, without having to trawl though individual census reports to find the subset of data that they are interested in. Secondly, there is the desirability of only having to enter a query at one portal, and for that query to then search all relevant metadata catalogues to find any data that match the query. The major obstacle to this is the development and use of standards that enable the portal sending out the query and the portals receiving it to understand one another. Gazetteers also have major potential here, as many textual databases may have large amounts of spatial information held within them, but have never been geo-referenced. An effective gazetteer would solve this problem by being able to convert the place names into spatial footprints, and then be able to investigate whether one place is, for example, within another or near another. This could also be done as part of a single query across the internet, to convert a place name given by the user at one portal into several different place names or spatial footprints from different gazetteers, which are then used to locate relevant datasets and/or records from yet more portals.

As has been stated several times in this book, the ability to integrate data through the use of location is one of the key advantages of GIS. Over the internet, this has the potential to reach its logical extreme. In theory, we are approaching a situation whereby a researcher can access a portal and enter a query to select all the information concerned with a particular place or geographical feature, or information near to the place or feature. The query assembles all of this using gazetteers and spatial metadata to search multiple websites for the information before assembling it and downloading it. There are still obstacles to be overcome in doing this, not least in the realm of intellectual property rights; however, the technology is nearing the ability to do this.

The effect of this on historical scholarship is yet to be seen, but it has the potential to make a huge impact that may be greater in history than in any other discipline. Fundamentally, the amount of primary information (in the historical rather than GIS sense) about a particular place at a particular time in the past is limited and, with the possible exception of new discoveries, does not grow. The challenge for historians is, and always has been, to make best use of these finite sources. Traditionally, this has meant that the historian spends large amounts of time in, and travelling to and from, archives and libraries. This is a highly inefficient way of working, and may result in sources in other locations being neglected due to the library's or archive's location. The potential for the technologies described in this chapter is for a query on the internet to be able to return all of the information relevant to the place, time and topic that the historian is interested in. This could then be downloaded to the researcher's desktop in a time that, compared to travelling to libraries, can be considered instantaneous.

The technological obstacles to this are being overcome; the remaining obstacles are more practical. A major constraint is in the creation of digital resources. These are clearly very expensive to create; however, significant investment is being put into constructing such resources in many countries, and this is far easier to justify if the data are to be effectively disseminated. There are also intellectual property rights issues. Here, history has a clear advantage over many other disciplines, as, while intellectual property rights issues need to be respected – and this will inevitably lead to constraints on access to data – historical sources do not have the same problems with copyright, commercial sensitivity and confidentiality constraints as modern data. Even when there are intellectual property rights issues, the internet can still be useful in allowing researchers to be made aware that the data exist and what the registration and/or purchasing requirements are.

Another issue is the acceptance of digital representations of sources. Many historians will be uncomfortable if they are not using the source directly, but are instead accessing a digital representation of it. To make matters worse, the digital product may not be a simple scan or transcription of the source, but may be a significant enhancement of the original, with all the potential for error and interpretation that this brings. Schaefer's (2004) Tithe Survey is a good example of this, as it brings together the tithe maps and schedule data to form a more usable product than the original, but one that may contain errors and inaccuracies. This is not a new issue. As was discussed in Chapter 5, historians have always made use of work of this type, albeit on paper, rather than in digital form. Historians treat this kind of work as works of scholarship that require scholarly acknowledgement and scholarly criticism. The role of metadata and documentation is vitally important in allowing this criticism to take place and the data to be used appropriately in the digital world.

A final criticism is that history based on digital sources in this way will neglect sources that have not been digitised. This is true, but has to be put in the context of the ability to search globally for digital sources, which is clearly more effective than being limited to sources that are available in a convenient archive.

8

GIS and quantitative spatial analysis

8.1 INTRODUCTION

As a tool, GIS is well suited to storing and manipulating large volumes of complex spatially referenced data. The challenge for the researcher is to turn these data into understanding and meaning. This usually involves simplifying the data sufficiently to make them comprehensible, while avoiding making them over-simplistic. Mapping is one approach to doing this, but, as Chapter 5 described, it has some important limitations. Spatial analysis is another approach that complements mapping. The spatial component of data in a GIS are fundamentally quantitative, as they consist of co-ordinate pairs – numbers that can be manipulated mathematically. Spatial analysis makes use of this to allow us to either summarise the patterns within spatial data, or to ask how attribute data are arranged in space. Having said this, the quantitative nature of spatial data does not mean that spatial analysis is necessarily positivist. In many ways, it can be the reverse, as rather than trying to produce aggregate summaries of an entire study area, it allows us to explore how different parts of the map behave differently. Therefore, rather than searching for similarities, much of spatial analysis is concerned with establishing how different places behave differently.

There are three broad approaches to spatial analysis. The first merely simplifies the map, using forms of smoothing to attempt to enhance the spatial patterns in the data. For example, smoothing a point pattern representing cases of a disease into a raster surface may help detect whether the pattern clusters, and if so, where. The second is concerned with asking whether a pattern or patterns that appear to the observer could simply have occurred at random. Finally, it may be that we want to investigate whether the clusters are associated with other factors or features such as factory chimneys, rivers, or population living in poor-quality housing.

Chapter 1 discussed the disciplinary and methodological origins of GIS. Given its scientific roots, it is no surprise that, traditionally, GIS has focused on statistical attribute data. It is a mistake, however, to think that quantitative spatial analysis is only a useful tool for quantitative studies of statistical data. Any study that wants to simplify spatial patterns to aid understanding of complex spatial data may find at least some of its tools useful. This chapter examines the advantages and limitations of using quantitative spatial analysis within GIS, and explores how it can be and has been applied to historical research. Section 8.2 defines spatial analysis and introduces its aims and philosophy. Section 8.3 looks at some of the problems of analysing quantitative geographical data. Section 8.4 explores some examples of spatial analysis techniques and how they have been applied in historical GIS. Within this section are sections on point pattern analysis, analysing point and polygon data with attributes, analysing networks and raster data.

8.2 WHAT IS SPATIAL ANALYSIS?

It is important to note that spatial analysis does not require GIS; in fact, it pre-dates it by several decades (see Gatrell, 1985). The advent of GIS, however, has led to a reinvigoration of the field, as the GIS data model – where attribute data are available – linked to a co-ordinate-based location is well suited to quantitative spatial analysis (Fotheringham *et al.*, 2000). Before the use of GIS became common among human geographers, Gatrell (1983: p. 2) defined spatial analysis as the study of three inter-related themes: *spatial arrangement*, which looks at the locational pattern of the objects under study; *space-time processes*, which concerns how spatial arrangements are modified by movement and interaction in space over time; and *spatial forecasting*, which seeks to forecast future spatial arrangements. Writing on the use of spatial analysis within GIS, Fischer *et al.* (1996: p. 5) follow a similar approach, but do not include spatial forecasting. At a low level, they define spatial analysis as being concerned with "[t]he use of quantitative (mainly statistical) procedures and techniques to analyse patterns of points, lines, areas and surfaces depicted on maps or defined by co-ordinates in two- or three-dimensional space'. At a higher level, they broaden this to state that spatial analysis is an approach that places emphasis 'on the indigenous features of geographical space, on spatial choice processes and on their implications for the spatio-temporal evolution of complex spatial systems'. This chapter follows this two-level definition. At the lower end, there is *spatial statistical analysis*, which consists of techniques which make explicit use of space as represented by the co-ordinate-based spatial data within GIS. These are part of a wider field, *spatial analysis*, which is concerned with gaining insight into spatial and spatio-temporal phenomena or processes by using a quantitative approach.

A key part of understanding what can be achieved by spatial analysis involves defining the likely limitations of the approach. One subject that spatial statistical techniques are well suited to in human geography is *exploratory data analysis*, which seeks to provide descriptive summaries, to identify the major features of interest from a data set, and to generate ideas for further investigation (Cox and Jones, 1981; Tukey, 1977). Spatial analysis research questions are often less suitable for *confirmatory analyses*, which attempt to test hypotheses and specify models. This is because when using spatial data it is usually easier to say that certain spatial patterns or relationships appear to exist, than it is to specify causal relationships. The exploratory nature of spatial analysis leads Openshaw (1991b: p. 394) to summarise its limitations as follows:

> even the most sophisticated spatial analysis procedure will probably not progress the user very far along the path of scientific understanding . . . The purpose of that analysis would typically be to develop insight and knowledge from any patterns and associations found in the data, which will either be useful in their own right or else provide a basis for further investigation at a later date using different, probably non-spatial and more micro-scale methods.

This means that, just like mapping, spatial analysis is more suited to describing patterns than explaining the processes that create them. It tells the researcher where something is happening (or perhaps not happening), but leaves him or her to explain why it is (or is not) happening there.

Openshaw and Clarke (1996) provide a list of basic rules that spatial statistical analysis techniques in a GIS environment should follow. The most relevant to historians include: firstly, that the technique should be study-region independent, which means that techniques that use statistics such as the average rate across the study area should be avoided; secondly, that the analysis should be sensitive to the special nature of spatial data, including issues such as spatial nonstationarity, spatial dependence, the modifiable areal unit problem and ecological fallacy – these will be discussed in detail below; thirdly, the results of the analysis should be mappable to allow the researcher to explore spatial variation and complexity.

8.3 WHY ARE SPATIAL DATA DIFFICULT TO ANALYSE?

It might be assumed that any statistical analysis of data carried out by geographers would be a spatial statistical analysis of one form or another. This is not, however, usually the case. Most statistical textbooks aimed at geography students – for example, Earickson and Harlin (1994), Ebdon (1985) and Shaw and Wheeler (1994) – focus on techniques, such as chi-squared tests, students t-tests, correlation and regression, that pay little, if any, attention to the impact of space. They are often inappropriate for GIS data for two reasons. Firstly, as was discussed in Chapters 1 and 6, much of geography is concerned with how processes and their impacts vary over space. Most statistical tests of this type do the opposite; they produce *global* summary statistics or models for the entire study area that are not allowed to vary from place to place. This can be termed *spatial stationarity*, as it assumes the relationship under study remains stationary over space. A spatial analysis should allow for the possibility that processes and their impacts will vary over the map. Doing this requires the use of *local* analysis techniques (Fotheringham, 1997) that allow summary statistics or models to vary according to location. These techniques are designed to explore *spatial nonstationarity*, and their results are usually presented in map form. Secondly, at a more practical level, the nature of much of the data themselves is also often not well suited to conventional forms of statistical analysis. Three reasons can be identified for this: *spatial dependence*, the *modifiable areal unit problem*, and *ecological fallacy*.

Many statistical techniques assume that the observations under study are independently random – in other words, that each observation will not affect or be affected by any other observation. Clearly, with geographical data, where an observation is likely to have an impact on its neighbours, this is invalid. To quote Tobler's First Law of geography, '[e]verything is related to everything else, but near things are more related than distant things' (Tobler, 1970: p. 236). This means that geographical data are spatially dependent because the value of each observation will be influenced by neighbouring values. This is both a problem and an opportunity. It is a problem because it invalidates an underlying assumption of many conventional statistical techniques, including most forms of regression. It is an opportunity because the impact that one place is having on neighbouring places is exactly the kind of issue that people using GIS should be interested in. Techniques to explore this usually test for *spatial autocorrelation*, the degree to which observations in a dataset correlate with other observations in the same dataset (Goodchild, 1987). There are a variety of statistical tests that explore this, including global techniques such as Moran's I and Geary's C that attempt to measure the degree of spatial autocorrelation across the entire dataset (Bailey and Gatrell, 1995; Cliff and Ord, 1973; Goodchild, 1987). More recently, local techniques, such as G_i and G_i^* (Getis and Ord, 1992 and 1996) and a variety of Local Indicators of Spatial Association (LISA), have been developed that explore where on a map values cluster or are dispersed (Anselin, 1995). Variations on LISA also allow users to explore how values in one dataset relate to values in another.

Many quantitative sources used by historians are published in aggregate form using arbitrarily defined spatial units such as registration districts, enumeration districts, census tracts or counties. The results of any analysis based on these data will be influenced by the arrangement of the boundaries of these units. This is referred to as the modifiable areal unit problem (Openshaw and Taylor, 1979; Openshaw, 1984; Fotheringham and Wong, 1991). It is caused by the combined influence of what are referred to as the *scale effect* and the *aggregation effect*. The scale effect means that, as data are aggregated, their values will become increasingly averaged. For example, two counties may appear to have average wealth, but it may be that when we disaggregate them we find that one has average wealth at tract level while the other has extremes of rich and poor which becomes average when aggregated together. The scale effect usually means that relationships appear to become stronger as data become increasingly aggregated. The aggregation or zoning effect refers to the impact different arrangements of aggregated areas have on the results of an analysis. If our rich and poor tracts are to be aggregated to form two new counties, different arrangements of boundaries may give either one rich and one poor county or two average counties, depending on which tracts are aggregated with which others. When this is done deliberately, it is called 'gerrymandering'; however, the aggregation effect means that all aggregate data are, to a certain extent, gerrymandered.

The issue of how to deal with the modifiable areal unit problem still causes controversy. On the one side, pragmatists such as Goodchild (1992b) argue that the spatial impact of the problem can be minimised by using data at the lowest possible level of aggregation. Others (see Openshaw, 1996; Openshaw and Rao 1995) claim that most existing techniques for analysing spatially aggregate data should be avoided, and call for new techniques and models to be devised that can not only cope with modifiable areas, but use them as part of an analysis. Others (for example Flowerdew *et al.*, 2001) devise statistical methodologies to help cope with the impact of modifiable areal units.

Ecological fallacy is related to the modifiable areal unit problem, but relates to the difference between aggregate patterns and those found at individual level. The terminology is confusing, as *ecological data* are data in aggregate form, such as those published in the census. If a researcher finds correlations in aggregate data, such as tracts, districts or counties, and assumes them to hold true at individual or household level, this is called ecological fallacy (Robson, 1969). Robinson (1950) provided an early demonstration of this by comparing aggregate correlations with individual experiences when analysing the relationship between ethnicity and illiteracy in the US. He demonstrated that there was an aggregate relationship between states with large proportions of black people and states with high rates of illiteracy, but showed that it did not hold that black people necessarily had higher illiteracy rates than other ethnic groups at an individual level. This was because the poorer states in the South spent less on education than richer states, leading to lower educational achievements. As the black population was concentrated in the South, this led to it appearing that there was a relationship between illiteracy and race. In actual fact, within southern states, no evidence was found that black people had higher rates of illiteracy than their white counterparts.

As with the modifiable areal unit problem, this is a fundamental issue inherent in polygon data that the researcher must be careful to manage. Some researchers have re-examined ecological fallacy in an attempt to create statistical models that will resolve the problem (Wrigley *et al.*, 1996). Techniques such as multi-level modelling bring data from different scales together, to build a clearer picture of these complex inter-relationships (see Bondi and Bradford, 1990; Jones 1991a and 1991b). Mathematical and statistical solutions will not, however, eradicate this problem entirely, especially in historical research where only limited data are likely to be available.

None of these issues is new; quantitative geographers have been aware of them for decades. The development of GIS, and particularly its spread into new areas within history and historical geography, together with an increasing availability of spatial data, means that we need to be conscious of their impact. GIS-based spatial analysis is opening up new approaches to deal with spatial nonstationarity and spatial dependence, which may make spatial analysis a much richer source. The modifiable areal unit problem and ecological fallacy, however, appear more intractable, and should perhaps be regarded as fundamental limitations of the source data, which require sensitive interpretation.

8.4 ANALYSING SPATIAL DATA

There are a growing number of spatial analysis techniques that can be used within a GIS environment. Commercially available software packages usually only have quite limited spatial analysis functionality. Instead, it is often preferable to use software such as GeoDa (Anselin *et al.*, 2006), CommonGIS (Andrienko and Andrienko, 2001) and STARS (Rey and Janikas, 2006). It is not the aim here to provide a complete description of spatial analysis approaches, as good textbooks are available (see, in particular, Bailey and Gatrell, 1995; Fotheringham *et al.*, 2000). Instead, we will provide a brief overview of pertinent techniques and, where appropriate, describe how they have been used by historians.

8.4.1 *Point pattern analysis*

The most basic form of spatial analysis is *point pattern analysis*, where the aim is to explore the spatial arrangement of a series of points. No attribute data are used; the aim is to summarise the spatial characteristics of the point pattern, to answer questions such as 'where are the points centred?', 'how dispersed is the pattern?', or 'are the points clustered, evenly distributed, or random?'

A historical example of a very simple descriptive analysis of point patterns is given by Longley *et al.* (2001). They cite a study that explores the changing patterns of land-use types in London, Ontario from 1850 to 1960, looking at four types of land use: residential, public and institutional, commercial and industrial. To do this, they make use of the *mean centre* and the *standard distance*, spatial equivalents of the mean and the standard deviation respectively. The mean centre is calculated like any other mean, by summing all of the x-values and all of the y-values and dividing them by the number of observations to give a mean (x,y) co-ordinate pair. The standard distance is the average distance from each observation to the mean centre (Fotheringham *et al.*, 2000). Formally, the mean centre $(\bar{\mu}_x, \bar{\mu}_y)$ is calculated as:

$$(\bar{\mu}_x, \bar{\mu}_y) = \left(\frac{\sum_n x_i}{n}, \frac{\sum_n y_i}{n} \right) \tag{8.1}$$

where (x_i,y_i) is the co-ordinate of the ith point and n is the number of points in each layer and the standard distance, d_s^2, and is calculated as:

$$d_s^2 = \frac{\sum_n (x_i - \bar{\mu}_x)^2 + (y_i - \bar{\mu}_y)^2}{n} \tag{8.2}$$

These two measures are used to give a diagrammatic summary of the changing patterns of these four classes of land use, as shown in Fig. 8.1. These maps show that, over time, the city has spread, and its land use has become increasingly segregated. The commercial district has remained in approximately the same location, demonstrated by the mean centre not moving, but its increased standard distance shows that it has expanded outwards. Industrial land use has moved eastward and expanded. Residential land use has always showed the most dispersed pattern, and its mean centre has remained relatively static, while its standard distance has increased. Public and institutional data has only been available from 1950 to 1960, but in this time it has moved northwards and become more dispersed.

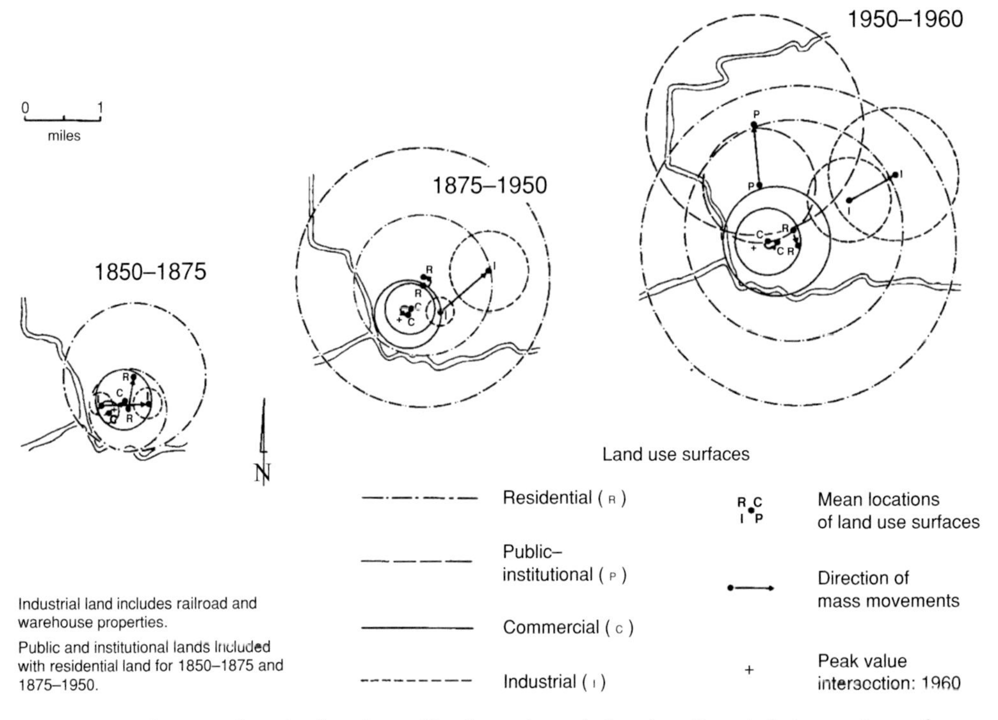

Fig. 8.1 Changing locations of land-use classes in London, Ontario between 1850 and 1960. The map makes use of the mean centre and the standard distance to summarise how land-use locations have changed over the period. Source: Longley *et al.*, 2001.

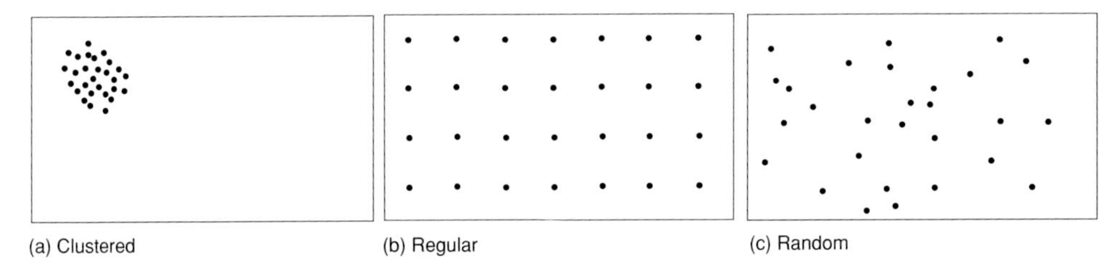

Fig. 8.2 Examples of point patterns showing (a) clustered, (b) regular and (c) random distributions.

This analysis uses simple statistics in a spatial context to summarise complex changing geographical patterns in a visually effective manner. The analysis could be performed without GIS; however, GIS allows easy manipulation of the potentially large numbers of co-ordinates, and it is straightforward to map the results.

A common aim of point pattern analysis is to test where the pattern of points exhibits clustering (as shown in Fig. 8.2a), shows a regular distribution where points

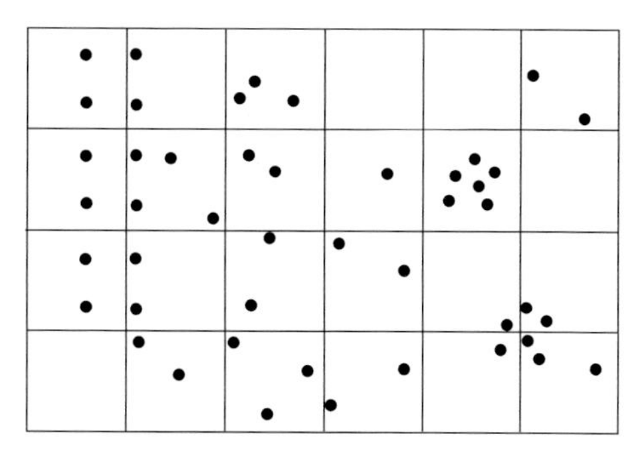

(a) A regular lattice placed over the points

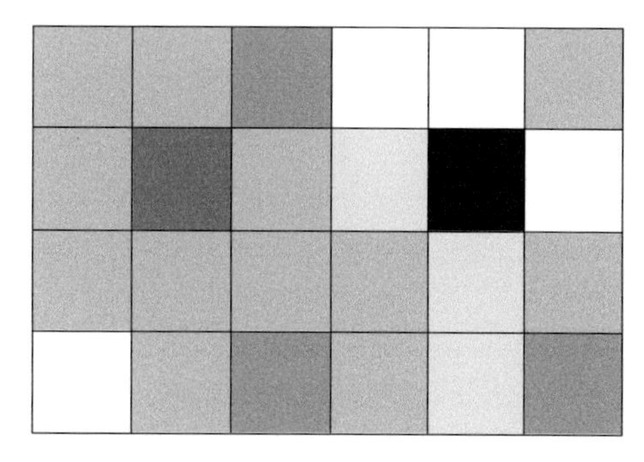

(b) Cells shaded using the number of observations

Fig. 8.3 Quadrat analysis. (a) A regular lattice is placed over the points and the number of observations in each cell is counted. (b) Each cell is shaded according to the number of observations it contains, with higher numbers of points having a darker shade.

are evenly spread out (Fig. 8.2b), or shows *complete spatial randomness* (CSR) where no pattern can be discerned (Fig. 8.2c). The challenge is to establish which of these is occurring, and, perhaps, where on the study area it is happening.

The simplest form of point pattern analysis involves the use of *quadrats.* The study area is subdivided using a regular lattice, such as a square grid, and the number of observations in each cell is counted. To explore the pattern produced by the quadrats, simple mapping can be used, with each cell shaded according to the number of observations it contains. In Fig. 8.3a, a set of points has been subdivided using quadrats. In Fig. 8.3b, the quadrats have been shaded according to the numbers of observations that they contain. If required, formal statistical techniques such as the *index of cluster size* (Bailey and Gatrell, 1995) can be used to conduct a confirmatory test for clustering or dispersion.

A major problem with using quadrats is that it imposes an arbitrary geography on the dataset, and, clearly, the results will depend on the size of the quadrats and where their boundaries fall. An alternative approach is to use *nearest neighbour analysis.* In its simplest form, this involves calculating the distance from each point to its nearest neighbour. The frequency distribution of these distances can then be examined. Short distances suggest a distribution that is heavily clustered, middle-sized differences imply a regular distribution, and where no clear pattern is shown, this implies randomness. Figure 8.4 shows the use of *box-and-whisker plots* to explore a variety of nearest neighbour distances. In each case, the 'box' in the centre of the diagram represents the inter-quartile nearest neighbour distance, with the line in the centre of the box representing the median. The 'whiskers' show the distances to the minimum and maximum values (Fotheringham *et al.*, 2000). These simple

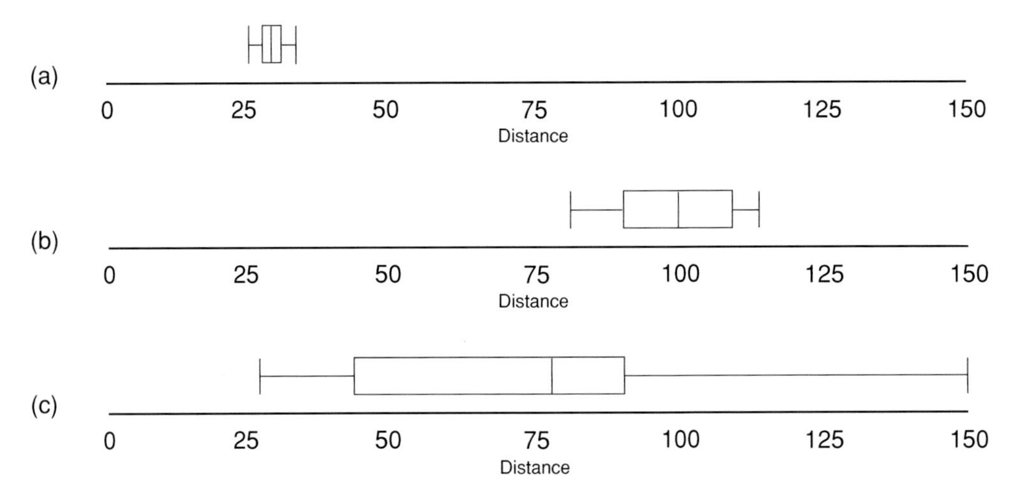

Fig. 8.4 A box-and-whisker plot to graphically represent the median, inter-quartile range, maximum and minimum. (a) and (b) give examples of patterns that might be expected from clustered and regularly distributed data respectively, while (c) uses the data from Fig. 8.3.

statistical summaries allow us to examine the pattern in an exploratory, non-spatial way. Figure 8.4a shows a highly clustered distribution, as all the points are close to their nearest neighbour. Figure 8.4b shows a regular distribution, as all of the points are a considerable distance from their nearest neighbours, and the nearest neighbour distances do not vary much. Figure 8.4c uses the data from Fig. 8.3. The pattern shows some evidence of clustering and some of dispersion. This correctly reflects the pattern, but is unable to say where each of these is occurring. Nearest neighbour analysis has a big advantage over quadrats in that no arbitrary units are imposed on the data. If required, nearest neighbour patterns can be mapped to show how they vary across the study area.

Other approaches dealing with point data include *density smoothing*, which turns a point layer into a raster layer reflecting the distribution of the points, *distance decay modelling*, and more advanced techniques such as Openshaw *et al.*'s (1987) Geographical Analysis Machine (GAM), which builds on the ideas above to explore a point pattern for local clustering against a varying background population.

8.4.2 *Spatial analysis of point and polygon data with attributes*

Spatial analysis can also be done with point or polygon data that include attributes. We will only brief touch on these techniques. Again, the basic idea of any spatial analysis is to explore how the data are arranged over a study area. The difficulty is how to model the relationship between a location and its surrounding observations. This is usually done using a *spatial proximity matrix*, which quantifies the relationship between each observation and each of its neighbours. Its values are referred to as w_{ij}, which is the weighting factor that each observation i has on a neighbouring value j. For points, this is usually based on the distances between a point and its neighbours. For polygon data, it is more difficult. One approach is to use the co-ordinate of the centroid of the polygon (see Chapter 4) and calculate values using distances in the same way as for points. An alternative approach involves using whether polygons share boundaries. Another method is to use the length of shared boundary as a proportion of each polygon's total perimeter. Examples of approaches are shown in Fig. 8.5. In Fig. 8.5a, centroids that lie within 3,000 m of each other have been given a w_{ij} of '1', and those that are further away are given a w_{ij} of '0'. This is somewhat unsatisfactory, as it relies on an arbitrary threshold that centroids either do or do not lie within. To avoid this, Fig. 8.5b uses a *distance decay model* in which the values of w_{ij} will decline as we move away from location i. In this case, an *inverse distance squared* weighting is used where values of w_{ij} will be '1' at point i and will decline exponentially with distance from it, without ever actually reaching '0'. In Fig. 8.5c, polygons that share a boundary have a w_{ij} of '1', and those that do not have a w_{ij} of '0'. Finally, in Fig. 8.5d, the length of shared boundary as a proportion of the polygon's perimeter is used. Unlike the other measures, this does not give a symmetrical matrix – for example, the boundary between *A* and *C* represents almost a third of *C*'s perimeter, but as *A* is a much larger polygon it only represents 6 per cent of its perimeter. It should be clear that creating a spatial proximity matrix involves arbitrary decisions, and that the results of any technique based on values of w_{ij} will depend on how these values were calculated. Spatial proximity matrices do, however, form a vital part of many types of spatial analysis technique.

There are both *uni-variate* and *multi-variate* techniques to explore spatial patterns in attribute data. The former are concerned with a single variable; the latter, two or more. The G_i measure of local spatial autocorrelation introduced above is one example of a uni-variate technique. It is calculated for every point or polygon i using the formula:

$$G_i = \frac{\sum_j w_{ij} x_j}{\sum_j x_j} \quad \text{where } j \neq i \tag{8.3}$$

Note that x_j refers to the attribute value of x rather than the x co-ordinate. The first stage in implementing this is that for each point or polygon i, the values of w_{ij} are calculated using an approach like the ones described above. These are then

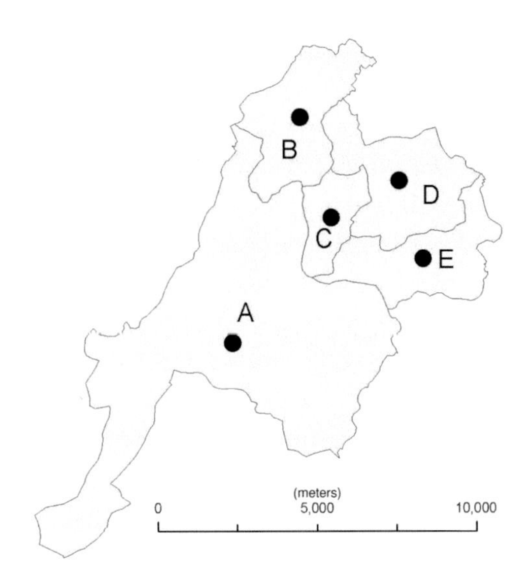

		A	B	C	D	E
	A		0	0	0	0
	B	0		1	1	1
i	C	0	1		1	1
	D	0	1	1		1
	E	0	1	1	1	

(a) Distance < 3,000 m

		A	B	C	D	E
	A		.019	.043	.019	.023
	B	.019		.096	.072	.029
i	C	.043	.096		.174	.090
	D	.019	.072	.174		.156
	E	.023	.029	.090	.156	

(b) Inverse distance weighting

		A	B	C	D	E
	A		1	1	0	1
	B	1		1	1	0
i	C	1	1		1	1
	D	0	1	1		1
	E	1	0	1	1	

(c) Share a boundary

		A	B	C	D	E
	A		.01	.06	0	.06
	B	.28		.06	.15	0
i	C	.30	.11		.38	.21
	D	0	.14	.20		.33
	E	.15	0	.10	.32	

(d) Shared boundary as a proportion of perimeter

Fig. 8.5 Examples of spatial proximity matrixes for five polygons. The dots are the centroids of the polygons. (a) Centroids less than 3,000 m apart have a w_{ij} of 1. (b) Inverse distance weighting using $w_{ij} = 1/d_{ij}^2$ where d_{ij} is the distance from point i to point j. (c) Polygons that share a boundary have a w_{ij} of 1. (d) w_{ij} is set by the proportion of shared perimeter.

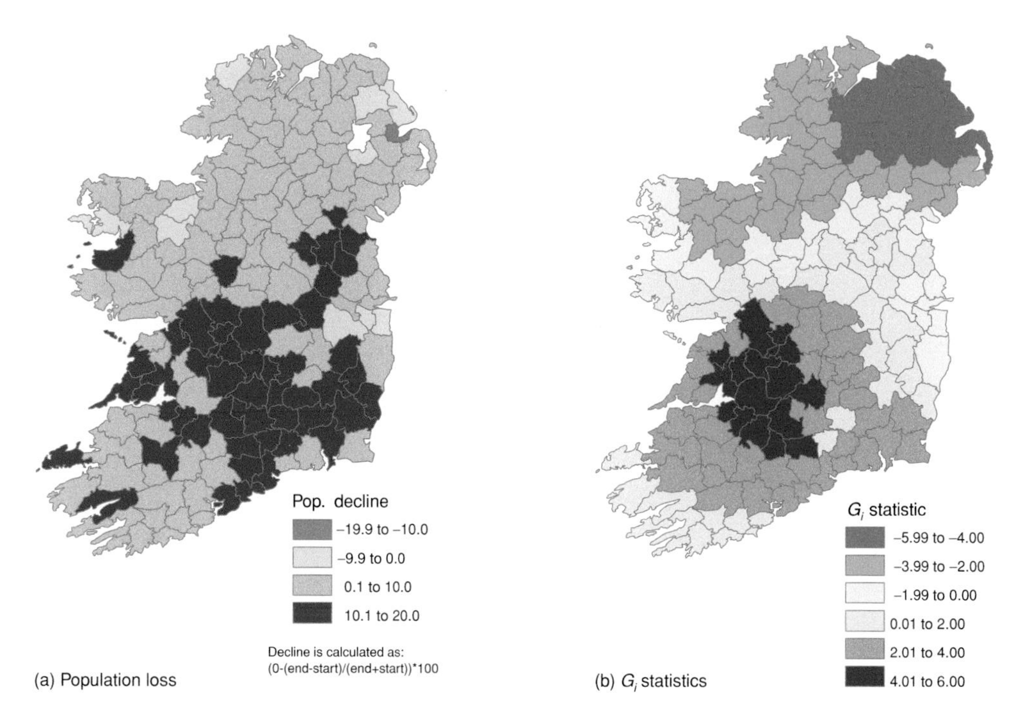

Fig. 8.6 G_i statistics of population loss following the Great Irish Famine. The Famine occurred in the late 1840s. (a) Population loss 1851–61 calculated as a normalised percentage: loss = 100 × (pop61 – pop51)/(pop61 + pop51). (b) G_i statistics of the pattern created using a distance decay model with a bandwidth of 70 km.

multiplied by the value of the attribute of each of the neighbours' x_j giving $w_{ij}x_j$. The sum of these for every neighbour is calculated and divided by the sum of the attribute values for every neighbour. High positive values of G_i show that a point or polygon is surrounded by high values, high negative values show that it has low values around it, and low values show that there is no clear pattern. The values of G_i are usually mapped to allow them to be explored.

Gregory and Ell (2005b) use G_i values to help analyse population loss following the Great Irish Famine of the late 1840s. Figure 8.6a shows the actual population loss from 1851 to 1861. In Fig. 8.6b, this has been smoothed using G_i statistics. This shows clearly that population loss was concentrated in the south Midlands and was lowest in the north-east of Ulster. This pattern is harder to discern from the raw data. This is thus a good example of a spatial statistical technique based on a spatial proximity matrix that uses a local and exploratory approach to simplify a map pattern in an attempt to summarise what it shows.

Geographically Weighted Regression (GWR) is a multi-variate technique that uses a form of spatial proximity matrix to perform a local version of regression analysis.

Table 8.1 *Global regression coefficients for Irish Famine regression. The dependant variables are small towns, large towns, proportion of fourth-class housing, and proportion of the population who were illiterate. All independent variables have been offset by their mean. Figures in bold are statistically significant at the 5 per cent level. The* r^2 *values are also shown*

	Intercept	S_town	L_town	Frth_clss	Illit	r^2 (%)
1840s	**13.5**	**−2.29**	**−12.99**	−0.01	**0.17**	38
1850s	**7.00**	1.51	−3.06	−0.04	**0.18**	10
1860s	**4.66**	−0.33	**−6.31**	−0.02	−0.04	15

In conventional regression, global summary coefficients are calculated such that the values in one variable, the *dependant variable*, can be estimated based on the values of one or more other *independent variables*. The regression coefficients are assumed to be constant across the study area. In GWR, a separate regression model is fitted for each location on the study area. This is done by modifying all of the independent variables using a spatial proximity matrix. This provides a descriptive technique that allows us to explore how relationships vary, and whether the coefficients show significant variations (see Brunsdon *et al.*, 1996; Fotheringham *et al.*, 2000; and Fotheringham *et al.*, 2002 for further details).

Gregory and Ell (2005b) demonstrate the use of GWR in their analysis of the Great Irish Famine. They first used areal interpolation to standardise all of their data onto a single set of Poor Law union boundaries, so that they could explore the impact of change over time without boundary changes influencing the results (see Chapter 6). They then wanted to examine the assumption that population loss was concentrated among the poor. There are no variables directly associated with poverty, so they used two variables from the census as surrogates: illiteracy and fourth-class housing. Fourth-class housing is the worst recorded by the Irish censuses of the time; it consisted of single-roomed dwellings with no windows, made of mud. They also added two dummy variables, to indicate whether the union contained a small town or a large town. These were set to '1' if the union contained a small or large town, and '0' otherwise.

The global regression parameters for this are shown in Table 8.1. All of the independent variables have been offset by their mean, which means that for the 1840s, for example, a Poor Law union with the average amount of fourth-class housing and illiteracy, and no town, lost 13.5 per cent of its population. Presence of a small town reduced this by 2.29 percentage points, and a large town by 12.99 points. Every percentage of the population who were illiterate over the average illiteracy would increase this by 0.17 percentage points, while below average would have the same effect in decreasing the population change. As the value for fourth-class housing is so small, and not statistically significant, it is not considered to have a noticeable impact on population change.

This last finding is particularly surprising, as the proportion of fourth-class housing declined from 36% in 1841 to 8.6% in 1861. It therefore seems odd that there appears to be no relationship between population loss over a decade and the amount of fourth-class housing at its start, but the pattern is consistent across all three decades. The GWR results, shown in Fig. 8.7, suggest a possible reason for this. It appears that in the east of the country, there is a positive relationship between the two variables, as would be expected. In the west, however, there is either no relationship or, in some places, the opposite relationship: high values of fourth-class housing seem to lead to relatively low values of population loss. Explaining this is difficult and would require further research. It does, however, provide further insight into the impact of fourth-class housing on population loss, and shows that the relationship between poverty and the impact of the Famine was more complex than global statistics would suggest.

A final use of GWR is shown in Fig. 8.8. Here a more complex GWR model has been created for population loss in the 1850s that has a total of nine independent variables. Rather than map the coefficients, Fig. 8.8 shows two measures of how effectively the model is working: local r^2 values and the number of variables that have a statistically significant *t*-value. Both are thus measures of how well the model is working at each location. Although there are some differences, the patterns between the two measures are broadly similar and show, in particular, that the model does not work well in the south of Ireland in the 1850s. This is an interesting finding, as it enables us to evaluate to what extent each place was 'typical' during the Famine. Skibbereen, in the south of Ireland, is often used as a case study; however, these results suggest that it is far from a typical place in the immediate post-Famine period. This, in turn, suggests that, perhaps, lessons learned from case studies on Skibbereen should be applied to other places with care.

8.4.3 *Spatial analysis and networks*

A final form of spatial analysis of vector data involves analysing distances across a network of lines. Networks can be analysed to find, for example, the shortest or lowest cost route between two or more points (know as the *travelling salesman problem*),

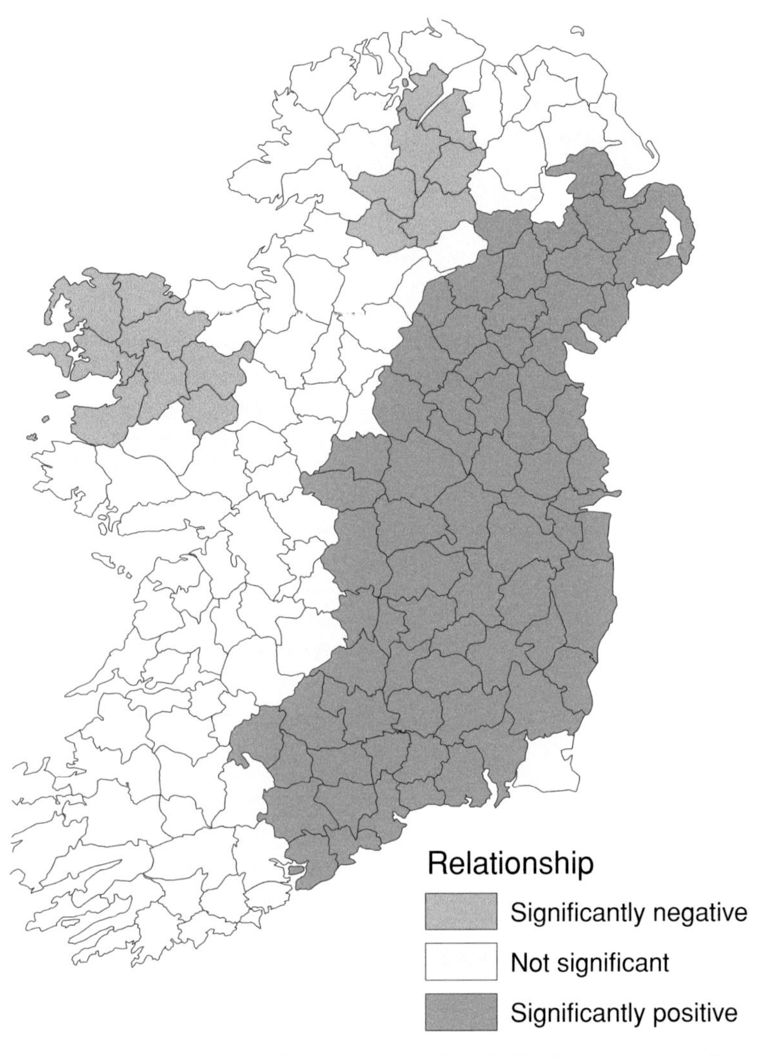

Fig. 8.7 GWR results for the impact of fourth-class housing on population loss in Ireland in the 1840s. The map shows the statistical significance of the GWR parameters for fourth-class housing.

or to determine the best location on a network to locate, for example, a factory, taking into account transport routes from raw materials (see, for example, Bailey and Gatrell, 1995; Heywood *et al.*, 2002; Verbyla, 2002). To date, no-one working in historical research has worked on this, but it is an area with much potential. A good example of a topic that this may help to shed new light on is industrial location, where detailed analysis of rail networks may provide new evidence about optimal locations for businesses.

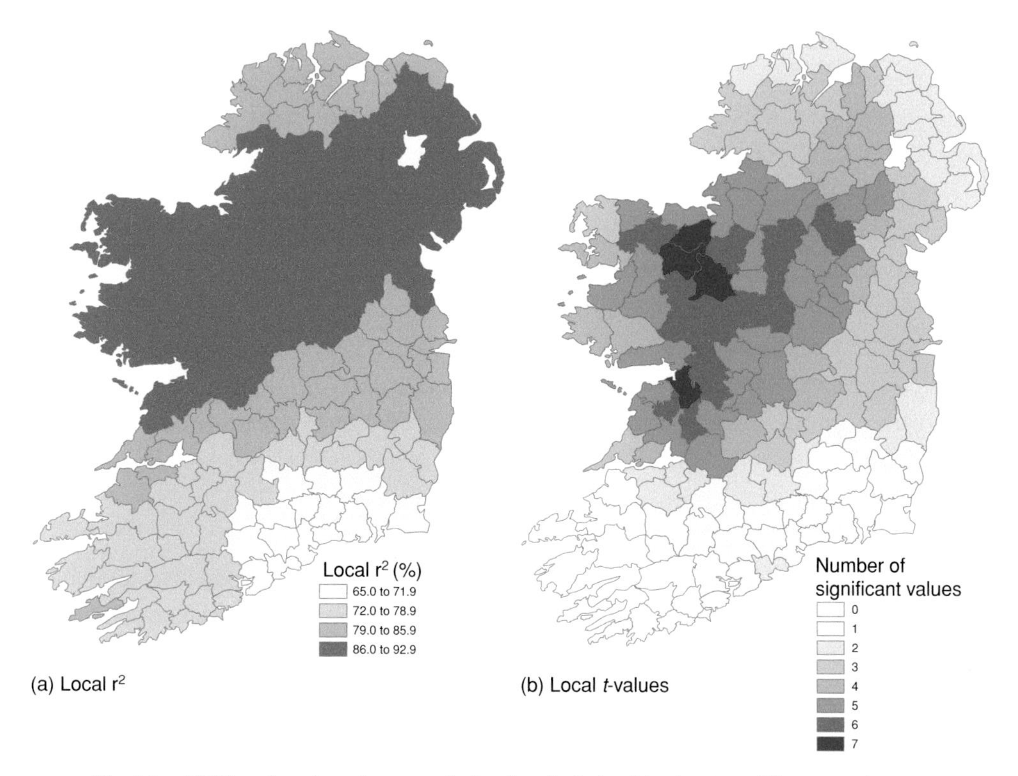

Fig. 8.8 GWR and explanation: population loss in Ireland in the 1850s. The maps show two possible ways of evaluating the effectiveness of a GWR model. In (a) the local r^2 coefficients are shown. In (b) the number of statistically significant t-values at each location is shown.

8.4.4 *Spatial analysis and raster data*

Spatial analysis can also be conducted with raster data. Sometimes the input layers will be raster data, but on other occasions it may be desirable to take point data and interpolate these to create a raster surface. This is a common approach in, for example, the oil industry, where drilling provides a number of sample points about the sub-surface geology from which they want to estimate a complete picture. It is also possible to take polygon data and convert it into a raster surface. This is usually done by first converting the polygons into points based on their centroids (see Chapter 4) and then converting these points to a raster surface.

The choice of whether it is better to analyse data as vector or raster very much depends on the nature of the data at hand. As was described in Chapter 2, vector is better suited to discrete features such as points or clearly defined areas, whereas raster is better suited for representing continuous surfaces. Raster data are also arguably better at handling uncertainty in data, as it does not require clear boundaries. Whether population is best represented using vector or raster is open to debate. As population data tend to be published using administrative units with hard, precise boundaries, they are often represented using polygon data, as these provide a good model of the way that the data are published. In reality, however, population does not change suddenly along administrative boundaries; indeed, many administrative boundaries are meaningless on the ground (Morphet, 1993). Equally, however, population does not form the smooth surfaces that raster data is most effective at showing. Bracken & Martin (see Bracken and Martin, 1989 and 1995; Bracken, 1994; Martin and Bracken, 1991) convert 1981 and 1991 Enumeration District-level (ED) data from the British census onto a raster grid to allow comparison, arguing that a raster surface gives a more realistic representation of the underlying population surface than polygons. Martin (1996b) describes a variety of exploratory data analysis techniques that can be used with the resulting raster surfaces.

Bracken and Martin's work on the 1981 and 1991 censuses is involved in re-modelling a very rich dataset, to take into account weaknesses in source's representation of space as a result of the use of administrative units. Bartley and Campbell (1997) provide another example of the use of raster data in analysis, but in a data-poor environment. They wanted to create a land-use map of pre-Black Death England, using the *Inquisitiones Post Mortem* (IPM), a source that provides a detailed breakdown of landowners' estates at the time of their death. The authors gathered a sample of 6,000 of these documents, covering the first half of the fourteenth century, and allocated them to point locations on a map of England. Each of these sample points contains information about the variety of land uses in the surrounding area. From the documents, they were able to identify twelve components of land use that they wanted to use as part of their analysis. The authors took their sample points and used them to create a continuous raster surface. This was done using a density smoothing method that calculated the value of each pixel based on all of the sample points within 250 miles of the cell. Nearer sample points were given more weight than further ones, using a distance-decay model. This allowed them to create a variety of raster surfaces. In one, shown in Fig. 8.9, they estimate the value of demesne meadow per acre. Their analysis was taken further by the use of cluster analysis to allocate each cell to one of six land-use classes: poor land with low unit land values, open country with little differentiation of unit land values, arable country with limited but valuable grassland, superior arable with pastures and wood, inferior arable and pasture with private hunting grounds, and open arable country with assorted lesser land values. To take into account the uncertainty in their model, the technique they used allowed cells that were hard to classify to be given second choice alternatives.

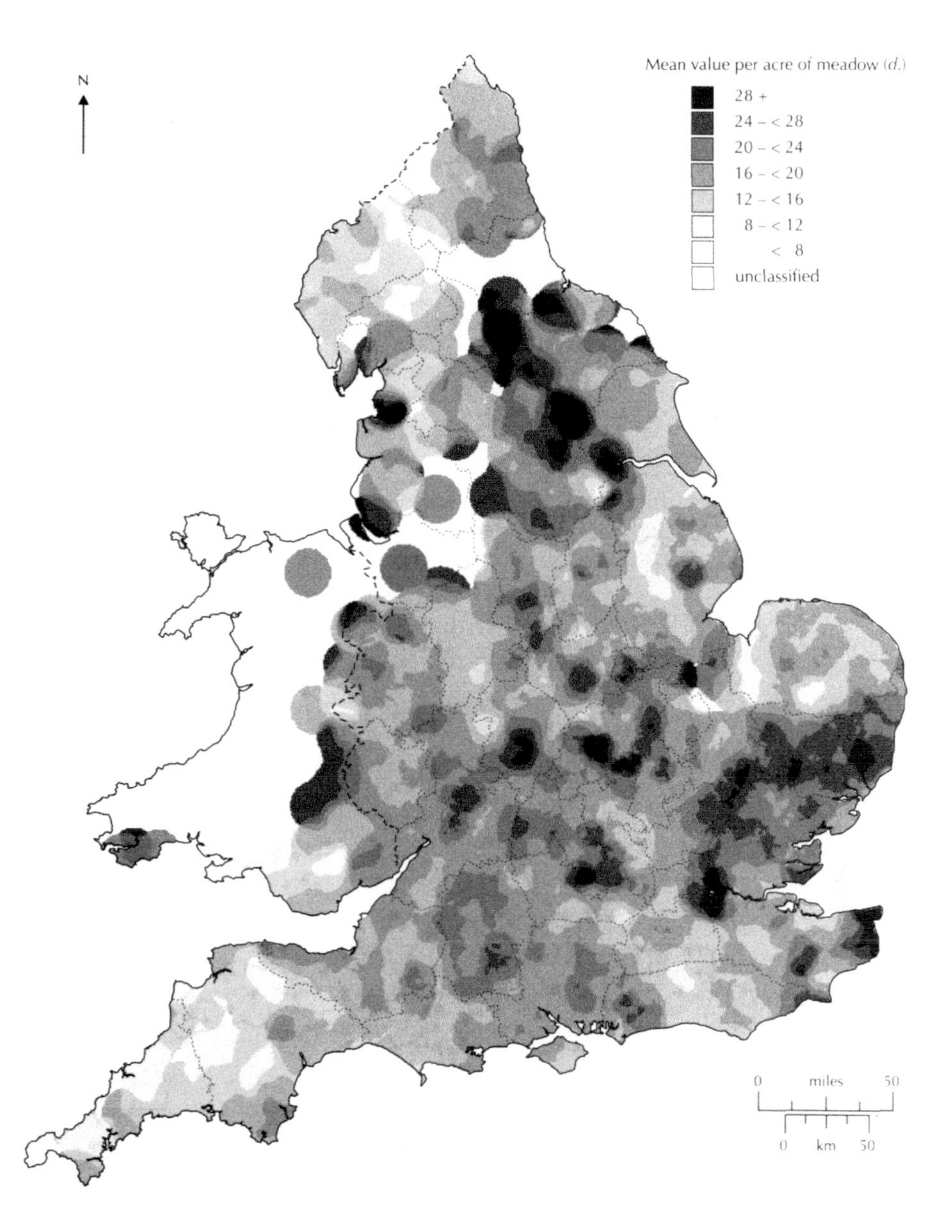

Fig. 8.9 Raster surface showing the value per acre of demesne meadow in pre-Black Death England. Values given in pence. *Source:* Bartley and Campbell, 1997.

In this approach, therefore, the authors have used the characteristics of a raster surface to allow them to handle the fact that their data are relatively poor. They have information about a large number of sample points and are able to use these to provide best estimates of the land use that surrounds these points. They have explicitly handled the uncertainty in their information by allowing cells to belong to more than one land use class. This demonstrates that GIS is not only useful in data-rich environments, but can also help to fill in gaps in relatively poor data, through the use of interpolation. Campbell (2000) goes on to use this data as a major component of his analysis of agriculture in medieval England, and Campbell and Bartley (2006) use it as part of an atlas of medieval England.

Bracken and Martin, and Bartley and Campbell, use relatively crude techniques to interpolate their data from points to a surface. There are more sophisticated techniques for doing this, such as *kriging*. This is described in detail in Bailey and Gatrell (1995), Burrough and McDonnell (1998), Fotheringham *et al.* (2000) and Isaaks and Srivastava (1989).

Raster data can also be used to explore multi-variate patterns. This is done by integrating multiple layers of raster data using *map algebra* or *cartographic modelling* as described in Chapter 4 (see also DeMers, 2002; Tomlin, 1991).

8.5 CONCLUSIONS: QUANTITATIVE SPATIAL ANALYSIS, GIS AND HISTORICAL GEOGRAPHY

A wide variety of spatial statistical techniques are available to the historical geographer wanting to analyse GIS datasets. The output from spatial statistical techniques can be in a variety of forms. Maps are very common, as many of the figures above show. On other occasions, it will be in the form of graphs, such as box-and-whisker plots. The output may also be in the form of a formal test statistic. Where local techniques are used, a formal test statistic may be available for every observation on the dataset, and these in turn are often best presented in the form of maps. Geographically Weighted Regression is an example of this.

The way in which space is modelled is crucial to most techniques. The choice of how to do this is often left to the user, but there are certain guidelines that should be followed. In particular, using arbitrary cut-offs, such as quadrats, is often less satisfactory than, for example, the use of nearest neighbours or distance-decay models. Analysing polygon datasets is hampered by the fact that we do not know how the population is distributed within the polygon. This means that we need to develop spatial proximity matrixes based on surrogates, such as centroids or adjacency, that may not model the polygons' characteristics particularly well. Whatever technique is used and however it is implemented, when interpreting the results of analysis it is vital that the researcher remains aware of the possible limitations of their data, the assumptions underlying the technique, and the impact of ways in which space is handled.

There is also much work that remains to be done before spatial analysis within GIS becomes judiciously and effectively used. Part of the reason for this is cultural:

even though the use of GIS is spreading in historical geography, the use of spatial analysis involves taking another significant and time-consuming step, where there is much for the researcher to learn. As outlined above, many techniques have been developed for spatial analysis within GIS; however, they are not commonly available within GIS software, and even if they were, they may still need modification to make them suitable for historical data. This means that this area of work is still very much at a pioneering stage.

In conducting a spatial or spatio-temporal analysis, the researcher should strive to explore a dataset in ways that allow a better understanding of the combined impact of the attribute, spatial and, where appropriate, temporal components of the data. This should allow a clearer grasp to be gained from the data than if space were ignored. There are, however, many limitations and challenges in the use of spatial analysis that the researcher needs to be aware of. The issue of spatial nonstationarity is a matter that needs to be addressed in any form of spatial analysis. It is rarely satisfactory to conduct a piece of geographical analysis that assumes that any relationships found will remain constant across the study area. Most statistical techniques, including some spatial statistical ones, do this. The researcher should be aware that spatial patterns and relationships will vary, and exploring these variations should be part of the aim of any analysis.

On its own, spatial analysis will rarely prove a causal link. In a classic early piece of spatial analytical research, John Snow mapped cases of an outbreak of cholera in Soho in London in 1854, and demonstrated that the outbreak was centred on a water pump (see Longley *et al.*, 2001). This appeared to provide evidence for his hypothesis that cholera was a water-borne disease. When the outbreak rapidly died out after the water supply from the pump was stopped, this provided further evidence for the link between polluted water and cholera. The spatial analysis demonstrated a possible link, but it took further non-spatial research – cutting off the water supply – to prove the association between cholera and polluted water. Therefore, a criticism that spatial analysis focuses on pattern, rather than process, may well be considered valid. This does not mean that spatial analysis is worthless. Both this study and the other examples given in this chapter, and the chapter that follows, do reveal spatial patterns that must be formed by a process or processes. Exploring these processes further will usually require other forms of work.

Spatial analysis is a tool for identifying patterns; explaining them usually requires further research. Equally, however, spatial analysis allows existing explanations to be challenged. It does this by exploring the extent to which the patterns that would be expected to be produced by the explanation actually do exist across a wide study area.

9

From techniques to knowledge: historical GIS in practice

9.1 INTRODUCTION

The previous chapters have largely talked about the theory of using GIS in historical research, focusing on the tools that GIS has to offer and how these can and should be implemented. In this final chapter, we explore how researchers have actually used it to better understand the historiographical topic that they are interested in. To a certain extent, this is a rather limited review because, although historical GIS has been around since the late 1990s, much of the early work was concerned with the development of databases, and more recently with techniques to exploit those databases. Only relatively recently have finished papers appeared, in which the results of a GIS-based analysis make a clear contribution to our knowledge of historical geography. The papers do, however, give a clear idea of how historical GIS can and is developing our knowledge of the past, and challenging orthodox historical explanations.

In Chapter 1, we made clear that GIS had its origins in quantitative Earth Science applications. When it spread into human geography in the early 1990s, it was still largely regarded as a quantitative technology, and indeed attracted much criticism for this (see, for example, Taylor, 1990 and the essays in Pickles, 1995b). While GIS is undoubtedly a good tool for performing certain types of quantitative geography, one of the most exciting developments in recent years has been the spread of GIS into qualitative geography, an area in which historical GIS has had much to offer. This is maybe the beginnings of Openshaw's (1991a) much maligned prediction that GIS would help to put 'Humpty-Dumpty', the discipline of geography, back together again, albeit perhaps in a far different form than he envisaged. At the same time, it is important to realise that there is far more to geographical scholarship than simply using GIS, which in reality is only a somewhat limited tool for storing and retrieving spatially-referenced data. Human geography, on the other hand, is a complex academic discipline concerned with the study of human populations at and between places (Johnston, 1983; see also Chapter 1).

This chapter is split into four broad sections: section 9.2 gives an overview of what researchers using GIS to study historical questions should be attempting to achieve, section 9.3 looks at how researchers using GIS to perform quantitative geography have approached the topic, and section 9.4 does the same for qualitative geography. Within this framework, three distinct approaches to GIS-based research can be identified. In the first, a large GIS database is built on the assumption that it will provide infrastructure for further research. In the second, GISs are created to answer a specific research question that challenges conventional assumptions, or to answer new questions in new ways. In the third, the data are explored to tell a story using a more narrative approach. These three approaches will be treated separately. The final section of this chapter provides a conclusion to the book that summarises what GIS has to offer to historical research, and identifies some key areas for future research.

9.2 APPROACHES TO TURNING SPATIAL DATA INTO GEOGRAPHICAL UNDERSTANDING

Although GIS stands for Geographical Information System, at a technical level, GISs are, in fact, really only spatial database-management systems. As described in earlier chapters, the data in a GIS has a co-ordinate-based spatial reference that gives it a location in space (see Chapter 2). Manipulating the data model also allows the data to be located in time (Chapter 6). A variety of basic tools, such as dissolving, buffering and overlay, can be used that allow the data to be manipulated in various ways through their spatial component (Chapter 4). The data can be visualised using maps and other types of scientific visualisation (Chapter 5), and various analytic techniques can be devised to make use of the fact that the data have a combined spatial and attribute (and maybe temporal) structure (Chapter 8). All of these are rudimentary tools that allow the data to be manipulated more effectively, mainly because all of the data are attached to one or more co-ordinate pairs.

There is a major difference between the definition of *spatial* used in GIS and the terms *spatial* or *geographical* used in academic geography. GIS contains spatial data, which is a crude quantitative representation of location based around Euclidean geometry and topology (Gatrell, 1983). Euclidean geometry gives an absolute location in space which allows the software to calculate straight-line distances between points and the angles between lines. Topology provides relative location by allowing the software to know which lines connect to each other, and which polygons and grid cells are adjacent to each other. GIS is thus concerned with precisely defined locations, the distances between them, and whether they connect. Geography, on the other hand, is about places. Defining place is more difficult. It is a complex cultural concept which is fundamentally concerned with human perception. Distances or connectivity are an important part of place, but, again, these concepts exist more in the human mind, rather than as precise physical characteristics. It is also a highly scale-dependant concept: a city may be a place, but the neighbourhoods that make up the city are all individual different and distinct places, and the streets or blocks that make up the neighbourhood are, again, new places (Cresswell, 2004).

The challenge, therefore, is how to turn spatially-referenced data into knowledge about places. The key advantage GIS has in doing this is that it is able to effectively subdivide a study area, the place under study, into multiple smaller places, and give some indications of how these places interact using spatial and perhaps temporal information. This is the key point about GIS; it contains information about many different locations. Any study should attempt to find broad trends within the places that these represent, and, perhaps, how the spaces interact over space, but also stress the diversity within and between places. In doing so, we actually stress geographical differences and similarities and avoid treating a study area as a homogeneous whole.

In analysing historical GIS data, the researcher always needs to be aware of the limitations of the source data. These limitations come in a number of forms. Firstly, all historical data are limited, and part of the historian's skill is to interpret the data accordingly. Secondly, there are the spatial and geographical aspects of the data, such as data quality (see Chapter 4), spatial dependence, the Modifiable Areal Unit Problem and ecological fallacy (see Chapter 8), as well as the broader aspects of simply considering what it means if places are close together or far apart. It is also important to consider whether close together in Euclidean or topological space is the same as being close together geographically, or whether physical or cultural barriers may mean that two places that seem to be near to each other may actually be quite separate. Finally, there is the easy-to-overlook point that GIS is a data-hungry technology. Where data are available in the correct form, GIS can be highly effective. Where there are no data, or the data do not fit neatly into the GIS data model, GIS cannot incorporate it. As a result, this information may be excluded from the analysis and also the understanding gained from the analysis. This means that the researcher has to be aware that this is happening, and be aware of non-GISable sources of Geographical Information.

9.3 GIS AND QUANTITATIVE HISTORY

9.3.1 National historical GISs

One of the first areas to receive attention in the field of historical GIS was the creation of national historical GISs. The Great Britain Historical GIS (Gregory *et al.*, 2002; Gregory, 2005) was one of the first of these, but many similar systems are either planned, being built, or completed. Good examples include the United States National Historical GIS (Fitch and Ruggles, 2003; McMaster and Noble, 2005), the Quantitative Databank of Belgian Territorial Structure (De Moor and Wiedemann, 2001) and the China Historical GIS (Berman, 2005; Bol and Ge, 2005). Gregory (2002a) and Knowles (2005b) provide reviews of national historical GISs. In general, these are a comprehensive database of the respective country's census and similar data, together with the changing boundaries of the administrative units used to publish these through the nineteenth and twentieth centuries.[1] As such, they are databases of most of the statistical information published about the country's society and how it has developed over the last two centuries. The GIS makes this an integrated database, as it takes statistical data, re-unites them with the boundaries for which they were published, and brings multiple census snapshots together. This gives a comprehensive representation of the source that no other depiction, be it on paper or in a database, can provide. It would seem, therefore, that this is an ideal structure with which to study trends over time and space in modern quantitative history. The reality, however, is that, although large amounts of time and money have been invested in these GISs, they are yet to make the kind of academic contribution that might be expected of them. This does not mean that they will not make this contribution, merely that it has taken longer than, perhaps, might have been expected. Using the example of the Great Britain Historical GIS, which the authors where heavily involved in, this section will discuss why this is the case.

The GBHGIS started in 1994 as a one-year project to add a 'mapping front-end' to an existing database of nineteenth-century statistics mainly concerned with Poor Law data published bi-annually (see Gilbert and Southall, 1996). To create this, a solution to the problem of researching changing boundaries and encoding them within a GIS had to be found. The data model used is described in Chapter 6. Once this had been devised, it was clear that its potential went far beyond a rather obscure set of nineteenth-century statistics, and that it had the potential to become the bedrock

[1] The China Historical GIS takes a rather different approach, as it is concerned with a far longer time period for data that do not refer to precisely bounded and well-mapped administrative units.

of a national historical GIS. Thus, sometime after the project had started, the Great Britain Historical GIS was first envisaged. Initially, it was felt that it was unlikely that sufficient funding would be provided by a single grant to build a system containing all the administrative units and associated attribute data that were considered desirable. Instead, progress was funded by a number of small- and medium-sized grants. Large-scale funding was finally raised in 1998 in the form of a £250,000 grant from the Economic and Social Research Council, which funded the project until the database was largely completed in 2000.

It thus took six years and over £500,000 from multiple grants to build the GBHGIS. Over this period, the project was not a conventional academic research project, which collects data, analyses it and publishes results. It was doing research, primarily on boundary changes, and encoding them in the GIS, but most of the challenges that faced the project were associated with project management, rather than the more traditional academic issues. Managing information technology projects is notoriously difficult, and this one was no exception, particularly as acquiring further funding was often a major task in itself, and satisfying the requirements of multiple grants whose aims sometimes conflicted with the overall aim of the project, did lead to problems. A further issue was that, in the UK, the bulk of credit for academic work is given for published outputs primarily in the form of peer-reviewed papers. A number of papers were produced that describe the database (Gregory and Southall, 1998; Gregory and Southall, 2000; Gregory *et al.*, 2002) and it seems ironic that more credit is given for these limited descriptions than for the resource itself.

The system was further extended under a National Lottery grant awarded in 2001, which gave the project £700,000. This had two aims: firstly, to extend the types of data that the system held, to include historical maps, place-name gazetteers, and qualitative data in the form of 'travellers' tales', contemporary descriptions of places which were geo-referenced using place names; and secondly, to put the resulting database online in an accessible way. This was launched in 2005 as the Vision of Britain website (Vision of Britain, 2005; Southall, 2006; see also Chapter 7). Again, this was not a conventional academic project, but was a database construction and dissemination project. Its major output is a database made available through a website. As was discussed in Chapter 7, whether this constitutes an academic publication is an interesting point. It is undoubtedly a major resource of widespread interest within and beyond the academic community. It is, however, data that are being disseminated, whereas in a conventional academic publication, analysis and interpretation would be required.

During the first phase of the project, some analytic work was done. Gregory (2000) discusses the use of the GBHGIS to study net migration in England and

Wales over the long term, and Gregory *et al.* (2001b) perform a study of long-run trends of poverty through the twentieth century (see Chapter 6). In both cases, the studies were more about potential and proof of concept than delivering a substantial contribution to knowledge on the two topics. They were limited by working with an incomplete database and also, as will be described below, by limitations of the techniques available to analyse these data. The work on inequality was taken further by Congdon *et al.* (2001) and Campos *et al.* (2004); however, both of these papers reverted to statistical approaches that made little use of the rich spatial content of the data. The statistical results in these papers refer to global summaries of the eleven standard regions of England and Wales, whereas the source data contained information on 635 registration districts. While these results may be interesting, this is far from the idea that a GIS-based analysis should make full use of the available spatial, temporal and attribute data.

As stated above, one of the key advantages of a national historical GIS is that it contains most or all of a country's census and related data in a single integrated structure. The main challenge in creating this structure is to unravel the boundary changes that affect the administrative units used to publish the data. The next stage for the GBHGIS, perhaps ironically, was to devise methodologies that would attempt to remove the impact of boundary changes. This required the development of areal interpolation techniques that enable long time series of data to be standardised on a single set of spatially detailed administrative units. These techniques are described in Chapter 6. Researching this showed that, while interpolating data crudely was easy, doing it in a way that minimised error was difficult, and different techniques were best for different variables. For example, different approaches are required for interpolating total population than when interpolating agricultural workers, a variable that follows a near inverse trend to total population. Developing methodologies to perform the interpolation effectively took more time (Gregory, 2002b; Gregory and Ell, 2005a). Even once these had been developed, it was apparent that any interpolation methodology inevitably introduced some error into the resulting data, so techniques had to be developed to identify where this error was occurring (Gregory and Ell, 2006).

Even given that appropriate techniques could be developed to interpolate all of the data in the GBHGIS onto a single administrative geography, this still leaves a resource that is still only a database. A further grant was required to help to develop appropriate spatial-analysis techniques for handling change over time and space within a GIS environment. As part of this, techniques such as Geographically Weighted Regression (Fotheringham *et al.*, 2002; see also Chapter 8) were applied to a pilot study. This study was concerned with population change after the Great Irish Famine in the mid-nineteenth century (Gregory and Ell, 2005b). The reason for using this topic, rather than one for Britain, was that it provided a research topic that had been extensively studied, but whose geographical elements were still poorly understood. It did so for a relatively small but self-contained area – Ireland, rather than Britain – and for a short period of time; only forty years. There is the potential for using the techniques that this work has piloted on analyses that cover two centuries of available data.

Therefore, in ten years of the project, there have been two distinct resource creation and dissemination phases, and two distinct phases of methodological innovation. While all of the significant challenges in these phases of the project have been largely overcome, it is still to deliver the substantial contributions to knowledge that it is capable of. These will follow in the near future. Although the investment that has been put into this may seem large, compared to the investment in collecting the statistics that they contain, these resources are small, and the potential for reinvigorating our use of these sources through this investment is large. This experience is typical of national historical GISs and other large GIS projects. The resource creation phases are very long, and recuperating these investments must also be seen as a long-term multi-stage process.

9.3.2 *Other quantitative approaches to historical GIS*

National historical GISs, therefore, spend large amounts of time and money creating resources in the belief that they will lead to substantial new research outcomes, even though the exact nature of these is not defined at the start of the research project. They are broad infrastructure projects. The alternative approach is to take an existing research question and add a GIS component to the study. These are more conventional academic research projects that may, or may not, create databases for use in future scholarship by other academics.

A good example of work conducted in a quantitative manner that uses GIS in this way is provide by Cunfer's (2002 and 2005) work on the Dust Bowl storms on the US Great Plains. The conventional explanation for the Dust Bowl was that it was caused by over-intensive agriculture by homesteaders who were swarming into the area at the time. This story originated amongst New Deal officials in the 1930s, and was perpetuated by a number of historical studies, especially that of Worcester (1982) whose argument was based on extensive case studies of two counties: Cimarron, Oklahoma and Haskell, Kansas. Using government documents, newspaper articles and interviews with residents, Cunfer set about testing this theory by using GIS techniques to explore all 280 counties in the Dust Bowl region.

The trade-off in Cunfer's analysis is that while he is able to get a far-wider spatial and temporal distribution than Worcester, the attribute data that he uses are relatively crude. He restricts these to county-level data, from sources such as agricultural censuses, soil surveys and weather stations. Sources such as newspapers are mainly used to give summaries of, for example, how often dust storms are mentioned in a county in a month or year. Much of his analysis was performed at an aggregate level; he was unable to comment on the behaviour of individual farmers.

The analytic techniques that Cunfer employs to explore his data are relatively simple, but nevertheless effective. He relies primarily on overlay, choropleth mapping, and makes some use of animation to look at spatio-temporal change (see Chapter 5). In its simplest form, his analysis simply involves comparing the major areas of dust storms in different years, with various possible causal variables, particularly soil type, cultivation, rainfall, and temperature. Rather than using statistical techniques, most of this work is based on comparing maps visually. This proves an effective way of comparing data from different sources over time and space. Comparing soil type with dust storms shows little evidence of an association between sandy soils – the most vulnerable to breaking up to form dust storms – and storm events. Comparing the area under the plough with dust storms reveals that some counties, including the two studied by Worcester, did dramatically increase their area under cultivation in the years immediately leading up to the peak of the dust storms. However, to complicate matters, other counties with extensive areas under cultivation did not experience storms, and some counties without much cultivation were within the peak storm area. As a result, the hypothesis that there is a link between ploughing and dust storms seems, at best, ambiguous. Two other variables that Cunfer uses are rainfall and temperature. These do show a strong spatial and temporal relationship with the areas most affected by storms. In particular, the areas most affected also had unusually warm and dry conditions at the times of the worst storms. Cunfer uses this to make an argument that the unusually warm and dry conditions in the mid-1930s were primarily responsible for the Dust Bowl, rather than insensitive agriculture driven by capitalism.

This analysis shows some of the strengths and weaknesses of a quantitative GIS-based study. The research is far broader in spatial and temporal scope than the traditional case studies whose findings it challenges. However, the thematic data that it uses are cruder, as it is limited to several key datasets at aggregate level. The study effectively challenges the conventional case-study-based explanation for the Dust Bowl. This is the strength of the geographical and temporal scope of the study. Where it is weaker is in its attempts to devise its own explanation. The relationship between dust storms and drought appears strong, but the processes by which this worked, the extent to which agriculture did contribute to this in some areas, and why farmers were unable to devise coping mechanisms to handle the storms, is less clear.

Hillier (2002 and 2003) performs another study that uses GIS to challenge historical orthodoxies on a different topic, at a different scale, and using different methods. Her topic is the practice of 'mortgage redlining', which she studies for the city of Philadelphia in the 1930s. In 1933, the Home Owners' Loan Corporation (HOLC) was set up to help both home owners and mortgage lenders who were struggling as a result of the Depression. Its role was to make low-interest loans, to cover defaulted mortgages. To assist in its provision of loans, the HOLC subdivided urban areas into 'residential security zones' that graded areas into four types, depending on the risk of lending in that area. The areas were graded from 'A', the best areas shaded in green, that were expected to be racially and ethnically homogeneous and have potential for residential growth, to 'D', shaded in red, that had low rates of home ownership, poor housing and an 'undesirable population or an infiltration of it', a phrase that was claimed to refer to the presence of Jews or African-Americans. In the 1970s, it was noticed that areas that had been graded 'D' in the 1930s had experienced massive under-investment over the ensuing forty years. This was thought to be due to a process known as 'redlining', whereby mortgage lenders would refuse to make loans to poorer areas and, in particular, areas with large African-American populations. Hillier argues that it had become an accepted fact that areas graded 'D' by the HOLC were perceived as too risky for mortgage lenders to lend to, and thus for builders and developers to build in. This all but guaranteed their long-term decline. She goes on to argue, however, that this logic had never previously been properly tested.

In order to test the theory, she performs a detailed analysis of the city of Philadelphia in which she asks three key questions: firstly, did areas graded 'D' have fewer loans made to them by the HOLC in the 1930s than other areas?; secondly, were areas with large African-American populations more likely to be graded 'D' than comparable areas with different ethnic populations?; and thirdly, in the years that followed the Depression, did areas identified as 'D' on the HOLC maps have either lower numbers of home loans or loans made at less favourable rates of interest than other areas.

To answer the first question, Hillier took a random sample of 300 loans and georeferenced the locations of the properties they were made on. These were overlaid onto a layer showing the residential security zones as polygons. The results showed that 62 per cent of the loans in the sample were made on properties in grade 'D' areas. This indicates that, contrary to popular belief, the HOLC was probably helping in areas of the greatest need, as it was supposed to. She then used tract-level data from the 1934 Real Property Survey, giving data on property and residents, and the 1940

Population Census, to see if areas with high African-American populations were more likely to be graded 'D'. Doing this involved using regression techniques that were able to handle the spatial autocorrelation inherent in the data, rather than conventional regression (Chapter 8). Her analysis confirmed that areas with high African-American and recent immigrant populations did indeed receive lower grades than other areas when other factors were controlled for. To test whether loans from private lenders were affected by areas identified as 'D' on the HOLC maps, Hillier, again, took a sample of loans from 1937 to 1950 and geo-referenced the locations of the properties that they referred to. She then tested both whether areas within grade 'D' zones, or near to their perimeter, had an effect on the number of loans, the loan-to-value ratio, or the interest rate. The results of this showed only that interest rates were slightly higher in grade 'D' areas.

Hillier's results suggest that areas with high African-American populations were more likely to be redlined than other areas, but contradict the commonly held belief that once an area fell within or close to the proximity of the redline, it was doomed to decline, as it would be impossible to get a mortgage in it. The results suggest that the practice of redlining was more complicated than had previously been assumed.

Like Cunfer, Hillier uses GIS to integrate data from a number of different sources: a sample of loans, the residential security zones, the Real Property Survey, and the 1940 census. While Cunfer relies heavily on comparing maps, Hillier uses overlay and statistical analysis techniques, although her papers are still well illustrated with maps. Both papers show that by using different approaches to integrating, analysing and visualising complex and disparate datasets, GIS is able to help undermine historical orthodoxies.

A different approach to performing quantitative historical GIS research is illustrated by Beveridge (2002). Cunfer and Hillier's approaches are effectively a broad form of hypothesis testing. Beveridge's approach is very different. He takes data on population growth, ethnicity, and various other social variables taken from the census, and uses these to tell the story, or rather stories, of population growth and ethnic divisions in New York City as it developed through the twentieth century.

Beveridge starts with a GIS of tract-level census data for New York from 1900 to 2000[2], and uses this to tell a quantitative narrative about how the population of New York grew over this period, and how this affected the diversity and concentration of various ethnic groups. The narrative is richly illustrated with maps, but is entirely descriptive. It shows, for example, how different ethnic groups had very different

[2] His data were taken from the US NHGIS, showing that it can be possible to work on a subset of NHGIS data while the national database construction process is still in progress.

settlement patterns in the early part of the century, and that these changed over time. It also describes the impact of the movement away from the urban core into the near suburbs and the periphery, and that wealth followed this movement. Choropleth maps and some statistical analysis are used, but the core of the research is a narrative about how the different neighbourhoods of the city developed differently with a particular emphasis on their ethnic make-up.

Unlike Cunfer or Hillier, this approach simply takes the data and uses it to tell a story of change over time and space. There are, however, similarities, and these are found in most pieces of quantitative GIS research. Firstly, the pieces of research are based on the available data. This limits the variables, or perhaps, more broadly, the topics, that they can study. Secondly, they are based on aggregate data and thus do not really tell the story of an individual's behaviour. Finally, both approaches are far better at describing the spatial and temporal patterns of the phenomenon under investigation than they are at explaining them. With Cunfer and Hillier, this means that they can challenge the existing theory effectively, but their explanations are more tentative. In Beveridge's case, he is able to say what happened, where and when, but does not attempt to unravel the complex processes driving these patterns.

All of the papers above approach places as being locations on the Earth's surface, and put this at the core of their analysis. In an ingenious analysis of fertility in twentieth-century China from the 1960s to the 1990s, Skinner *et al.* (2000) take a very different approach to defining place. Rather than use physical location, they attempt to locate the different parts of China into what they term *Hierarchical Regional Space* (HRS). This is a matrix that allocates every location into a cell, determined by how urban or rural it is on one axis, and how core or peripheral it is on the other. Places are classed from inner-core urban to far-periphery rural. The paper, therefore, focuses on a conceptual model of geography that puts places into a matrix that represents how urban and how peripheral places are, and uses this to structure the analysis.

They use three main sources of data to create the HRS matrix: 12 million records from the 1 per cent sample of the 1990 Census of China, the boundaries of 2,800 administrative units, and the locations of 12,000 towns and cities. To create the typology of places, they draw on the ideas of central-place theory, that provides a theoretical background to a hierarchy of settlement size, and core-periphery models to give a structure to the relationship between towns and their hinterlands. They subdivide China into an eight-level urban-rural hierarchy using a combination of information, including settlement sizes and variables on industrial structure and health and education facilities. They create their seven-level core-periphery hierarchy using a variety of socio-economic indicators on subjects such as literacy rates, population structure, employment indicators and levels of agricultural activity.

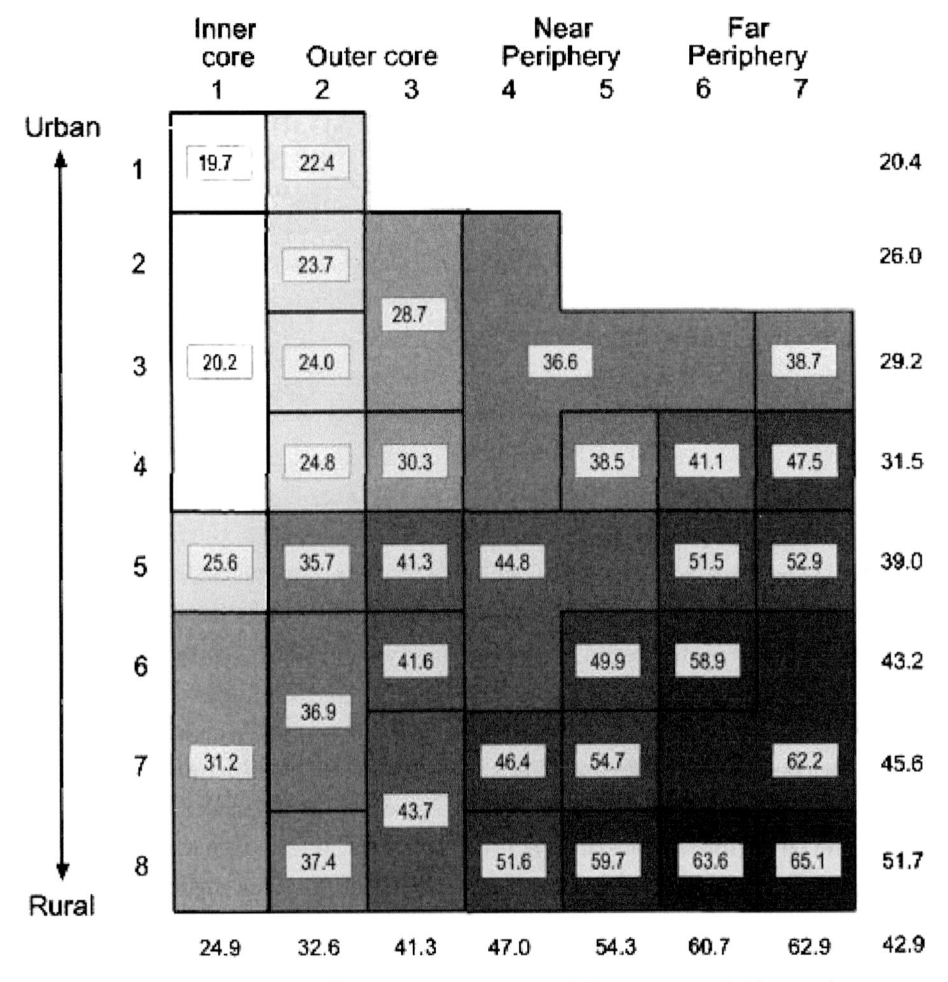

Fig. 9.1 The percentage of women aged over 30 with three or more children in the Lower Yangzi region of China, differentiated using a hierarchical regional structure. Source: Skinner *et al.*, 2000.

Combining the urban-rural hierarchy with the core-periphery hierarchy gives a two-dimensional matrix of hierarchical regional structure in which the individual-level records from the 1 per cent census sample could be located. The top-left hand corner of the matrix corresponds to inner-core urban areas, while the bottom-right hand corner shows far-periphery, rural areas. Figure 9.1 shows an example of how this matrix can be used, showing the percentage of women aged over 30 with three or more children.

Once they had simplified the geography of China into an HRS matrix, they were able to examine spatio-temporal trends in different types of areas, based primarily on time-series graphs rather than maps. A conventional analysis might be expected to contrast how place *A* differs from place *B*. In this analysis, however, Skinner *et al.* contrast how, for example, the inner-core urban places differ from the far-periphery rural. They show that, as we move from the inner-core urban to far-periphery rural places, the date at which fertility decline started becomes later and later.

Their idea of space is, therefore, not conventional co-ordinate-based space, but is instead hierarchical regional space, defined by the socio-economic characteristics of each place. From here, the analysis is no longer concerned with individual places, but is instead concerned with how different types of places, as defined by the HRS, behave differently. This is, therefore, an ingenious use of geography that goes well beyond the spatial.

9.4 GIS AND QUALITATIVE HISTORY

A recent and exciting development in historical GIS has been the increasing use of qualitative data within a GIS framework. Two projects in particular have been successful in creating and beginning to analyse complex qualitative historical GISs. The first of these is the Salem Witchcraft project which studies the famous witchcraft trials in Salem, Massachusetts in 1692 (Ray, 2002b). The second is the Valley of the Shadow project which is concerned with studying the differences between two counties, one North and one South, before, during and after the American Civil War (Sheehan-Dean, 2002; Thomas and Ayers, 2003). In both cases, the researchers have started projects concerned with collecting and analysing a large number of documentary sources related to their subject. The GIS component was added later due to its ability to structure the data according to location, and its ability to summarise it through maps. This, in turn, has led to advances in our knowledge of the subject areas.

Ray's project is concerned with a relatively small area – the village of Salem and its surrounds – over a short period of time. His data are concerned with events that evolved on a daily basis, namely the spread of accusations of witchcraft. They also have considerable spatial complexity, as he is interested in studying events at household level. The use of GIS, along with other database technologies, allows him to glean new knowledge from existing and well-studied sources.

The Salem witchcraft trials took place in the village of Salem in the Massachusetts Bay colony (now Danvers, Massachusetts) in 1692. In all, 168 people were accused of witchcraft in trials running from February until November. This led to nineteen people being hanged, one being crushed to death, and five dying in prison. It is one of the most studied events in American history. Ray started with a project to create a digital archive of sources relating to the trials: the *Salem Witch Trials Documentary Archive and Transcription Project.*[3] This includes complete transcriptions of contemporary court and other documents, transcriptions of rare books written about the trials, modern and historical maps of the village, as well as catalogue information from a wide variety of disparate archives. It has grown to contain over 850 primary source materials.

Over 300 people are named in the archive's documents as accusers, accused, or defenders of the accused. Using GIS allowed Ray to locate these people and the events that affected them in space and time. He did this using a map produced by a local historian, Charles Upham, in the nineteenth century that maps the locations of all of the households in the village. The GIS database that he created based on this map can be presented to users through queriable maps and animations. It also enabled him to explore the ages and sexes of the people concerned, the frequencies of accusations made against them, family relationships, and information on wealth and social status.

This allowed Ray to explore how the patterns that this revealed compared with explanations provided by earlier researchers. In particular, he investigated three approaches to understanding the trials. Firstly, he explored Charles Upham's hypothesis that the accusations were related to property disputes and personal grudges. Secondly, he explored Boyer and Nissenbaum's (1974) argument, advanced in the 1970s using Upham's map, that the village was split by social and economic pressures. They had argued that people living nearer the town of Salem were more involved in mercantile and political activities associated with the town, while those further away wanted political and religious independence from it. This manifested itself in the formation of a new church in the village and the appointment and subsequent dismissals of several ministers, including the Reverend Samuel Parris, the minister at the time of the trials whose salary was being withheld as a result of this dispute. Boyer and Nissenbaum summarised their argument by using a straight line running north to south on the map. They argued that most of the accused and their defenders lie to the east of this line, near to Salem town, while the accusers are mainly to the west of the line. This line, therefore, was used to summarise the split with those on the side of the line furthest from the town using allegations of witchcraft to further their desire for independence, and those nearer the town, who opposed further independence, being victims of the allegations.

The third is a hypothesis, put forward by Miller (1953), which looked at broader problems within the Massachusetts Bay Colony. These were related to a power

[3] http://etext.virginia.edu/salem/witchcraft. Viewed on 21 June 2007.

vacuum caused by the suspension of the colony's charter in 1690, and the failure to replace it with a new charter and governor until 1692. This led to a destabilisation of the colony's religious and legal governance which in turn led to personal conflicts running out of control.

Ray's mapping of Salem in itself throws doubt on Boyer and Nissenbaum's straight-line split. Firstly, Ray adds additional characters to the map, which blurs the simple east–west divide. Ray was also able to map the location of wealth throughout the village, using taxation records, and whether people attended the new village church and supported its minister. These patterns do not show a clear split across the village, allowing Ray to question the assumptions underlying Boyer and Nissenbaum's theory that the spatial patterns of accused and accusers was a surface manifestation of underlying social and religious splits within the village.

Ray also maps the broader patterns of accusations in the east of Massachusetts Bay that occurred at the same time. Many other towns also had witchcraft accusations. In some cases, these were directly linked to Salem. In Andover, for example, a round of accusations started when two girls from Salem were invited to the town to identify witches. Ray uses this to argue in support of the theory that there were broad problems within Massachusetts Bay that led to personal conflicts running out of control. While these were at their worst in Salem, the pattern was much broader than this.

Ray uses GIS as an extension of his existing approach. He did not start by conducting a GIS project, but the GIS has increasingly driven his analysis. It allows him to locate people in time and space, and to explore changing spatial patterns through a variety of attributes. It is this ability to handle multiple attributes, such as taxation records and church membership, that enables him to present a serious challenge to Boyer and Nissenbaum's argument. He is also able to produce maps at a broader scale that enable him to see the patterns in the area surrounding Salem, and thus explore how the detailed, household-level patterns of accusations within Salem related to broader village-level patterns within the colony as a whole. In this way, he is able to support arguments that the broad problems caused by a colony-level power vacuum exacerbated problems at the individual level. The historiography that Ray challenges was essentially spatial. Through his use of GIS, Ray is the first researcher who has had the tools at his disposal to rigorously investigate the spatial elements of the problem, and this is what enables him to challenge the earlier work.

The Valley of the Shadow is a larger scale project that follows many of the same approaches as the Salem Witchcraft project. It is also focused on a well-studied aspect of American history: the impact of slavery on the underlying causes of the Civil War. Again, the project is interested in studying how household-level behaviour impacts on broader trends. It does this by creating detailed archives of two counties on the eve of the Civil War: Franklin County, Pennsylvania, to the north of the Mason-Dixon Line; and Augusta County, Virginia, to the south. The aim of the project is to compare the two communities, not as counties as a whole, but rather by looking in detail at the individuals and communities that make them up. Both counties were physically close together and had similar topographies. Both were relatively prosperous and had successful and diverse economies. The main difference between the two was that slavery was all-pervasive throughout Augusta.

The project wanted to explore a basic paradox in our understanding of the causes of the Civil War. It is commonly argued that slavery made a profound difference between the North and the South. The North is characterised as being a modern democratic society with economic innovation and social mobility, and the South as being desperate to resist these trends. The problem is that studies have repeatedly shown little difference between the two in terms of voting patterns, wealth distribution, occupational levels, and so on. This leads to questions of whether there really were fundamental differences that made war inevitable.

Their database was based on sources including personal manuscripts, such as diaries and letters, the 1860 census returns, tax lists, and so on. As with Ray's project, they started to create a non-spatial database. A GIS became necessary because many of the diaries and other sources concentrated on local issues, so locating the source into its spatial context made them far more comprehensible. They therefore digitised detailed antebellum maps of the two counties, and used these to geo-reference much of their information, including places named in diaries, and household-level information from the census, returns. They were also able to integrate their data with modern data on soils and relief, which are important for understanding patterns of agriculture.

They find that both counties were wealthy and prosperous, and that both had thriving industrial and trade networks. White Augustan farmers were wealthier and had, on average, larger farms than their northern neighbours. Interestingly, they also show that free blacks in Augusta were richer than blacks in Franklin. This, they argue, showed that Augustan society believed itself to benefit whites, free blacks and slaves. They also show that the distribution of wealth was quite different between the two counties. In Augusta, wealth was concentrated in towns, while in Franklin it was found in rural areas. Augusta had more towns than Franklin, but these were smaller. Agriculture in Franklin was concentrated on wheat; in Augusta it was concentrated on corn which fed slaves and provided a raw material for the many distilleries in the county. This difference was, however, narrowing as Augustan farmers increasingly moved into wheat production.

As well as locating data in its place, GISs ability to integrate data also helps some of their analyses. One question that they are keen to explore is whether southern farmers were more isolated in rural areas than their northern counterparts. Superficially, this appears to be the case, as Franklin has a higher population density and a longer length of major roads. What the Valley of the Shadow project is able to show, however, is that Augusta has a much better developed network of minor roads, and this, combined with the relatively large number of towns in Augusta, leads to serious questions about whether southern farmers were as isolated as previously proposed. To do this, they used the GIS's ability to buffer data and to bring together data on roads, topography and settlement patterns.

The main argument that they advance is that the war was not an inevitable conflict between a modernising North and a South that was diametrically opposed to this, but was in fact based on different views of modernity. Their work is interesting in that they fuse qualitative and quantitative sources and techniques, demonstrating that GIS has the ability not only to integrate data, but also to integrate approaches. As with Ray, the Valley of the Shadow project is able to use a large qualitative archive to explore well-studied historical questions using GIS techniques. This enables them to answer detailed spatial questions that question assumptions made by previous studies.

9.5 CONCLUSIONS: TOWARDS A FUTURE FOR HISTORICAL GIS

In this book, we have discussed in detail the technologies, methodologies and scholarship of using GIS to study the past. While GIS is well established in many disciplinary areas, it is clear that its use in historical geography is still in its early stages. Indeed, a sceptic may argue that, to date, GIS has done little to justify the investment and attention that it has received. As described above, in the UK, North America, China, and Western Europe there have been very significant sums of money invested by research funders in the development of national historical GIS. These projects are, by their very nature, large and consequently costly, with budgets that may run into millions. The returns on these investments in terms of scholarly research publications are, so far, limited. It has been clearly established that the bulk of the expenditure on GIS projects in any discipline takes place early in the project cycle, whereas the bulk of the benefits do not arrive until later. This is precisely because building databases is such an expensive task but, once complete, these databases allow the sources that they represent to be explored in entirely new ways. As a result, the majority of new knowledge that has been derived from historical GIS projects to date has come from small- and medium-sized projects with a more limited lead-in time, such as Cunfer's study of the Dust Bowl, Hillier's examination of mortgage redlining, and Ray's study of the Salem witchcraft trials. These are three projects that take very different approaches to studying very different subjects. In all cases, however, they shed new light on the topic by their emphasis on the geographical patterns that they find in their data. It is the ability of GIS to structure, integrate, visualise and analyse these data that enables the authors to develop new understandings in research areas that had already been the subject of detailed debate by historians. We believe that historical GIS is currently at a stage of development where it can show what it can potentially offer to historical research, but where it still has much to deliver before it proves that it will become an integral part of the discipline of history. In this final section, we review the emerging trends within historical GIS, and how these are likely to impact upon it.

At its core, GIS is a technology, a form of software that adds spatial functionality to a database management system. GIS software is now relatively cheap, reasonably easy to use, and runs on standard PCs. Desktop GIS is thus well established and mature, and is likely to change little in the near future. Scholars can, therefore, proceed to use GIS tools in the confidence that the software is robust, well supported, and will not be transitory.

Similarly, key strategic resources are now available. In the past, research projects have often involved a significant element of data capture. Funders rightly saw that many research projects were gathering similar sorts of data and, in many cases, exactly the same information. This was particularly the case for what might be termed core or strategic data sources, such as historical censuses. Part of the rational for funding expensive historical GIS projects was to make these key datasets freely available. In the UK, virtually all published British census data from the first census in 1801, together with the reports of the *Registrar General*, are now available in digital form within the Great Britain Historical GIS. In the US and elsewhere, the situation is similar. This means that projects that would previously have had to include expensive and time-consuming data capture phases can now begin at the analysis phase. This not only reduces the cost of such projects, but also greatly enhances their attractiveness to researchers, as much of the tedious work for which there is relatively little credit has already been done. In addition, the data are not simply transcriptions of the original sources. By bringing together spatial and attribute data, a national historical GIS creates a representation that is closer to the underlying population than it has been possible to produce since the data were originally enumerated. In other words, much of the hard work has already been done, and, through documentation and metadata, the scholarly processes developing a national historical GIS are fully detailed.

These two key areas of technological development – software and resource creation – are thus well established. By contrast, a field that is beginning to evolve, and clearly has much potential in the humanities and social sciences, is e-Science and Grid technology. As Chapter 7 described, a key element of e-Science and the data grid is providing robust resource discovery tools concerned with integrating disparate resources over the internet. We believe that GIS has much to offer to this. One of the most effective ways of integrating apparently incompatible resources is through place. Virtually all humanities and social science data resources can be referenced by geographical location. As a consequence, these sources can be placed within a GIS framework. In most cases, this will not require the complex work of creating a polygon-based national historical GIS. Instead, it can be achieved rapidly through the use of place-name gazetteers which give a point location to every place that occurs in a source. As these gazetteers are also internet resources, the potential to integrate and explore geographical patterns in sources collected or published for very different purposes and stored in very different locations becomes enormous. Theoretically, it will be possible to begin to assemble all of the data about a place, available in digital form anywhere in the world, using a single internet query. This will give us unparalleled access to information about what a place was like in the past.

With or without e-Science, one of the major challenges with GIS, as with other technologies, is to devise methodologies that take the data held within the technology and convert it into information that is of use to the researcher. There are a core set of tools that lie behind many of the methodologies offered by GIS. These include the ability to query data, measure distances and angles, manipulate data using their spatial component through operations such as dissolve and buffering, integrate data through overlay techniques, and map patterns in a variety of ways. These procedures were described in Chapters 4 and 5. They form the core of more complex methodologies in fields such as visualisation and spatial statistics. As with the technology, many of the fields driving the developments of these methods are outside the discipline of history, so part of the challenge is to apply them appropriately within historical research. The other part of the challenge is to devise techniques that fill gaps that are not developed in other disciplines.

To date, the methodological developments specifically within historical GIS have been limited. The sole area where significant progress has been made is in the use of areal interpolation to create consistent time series of polygon data (see Chapter 6). In most other areas, historians have been content to use the methodologies handed down to them by other disciplines. There are two areas in particular where historians need to develop better methodologies. The first of these is approaches to exploring spatio-temporal change using both visualisation and statistical approaches. Although progress has been made with animations (see Chapter 5), this is limited, and statistical methodologies for handling spatio-temporal change are almost non-existent. As much historical geography is concerned with change over time methods that summarise spatio-temporal data in a way that identify broad patterns, how these vary over space and time, and where there are exceptions, are clearly required. The second area is the handling of error and uncertainty. As has been discussed in earlier chapters, this can be handled as a documentation or metadata issue, a data model issue through the use of raster surfaces to model point data, a statistical issue through the use of techniques such as fuzzy logic, or an interpretative issue. As one might expect, none of these approaches offers a complete answer. There is, however, a pressing need for historians to take a clear look at how the discipline has traditionally handled uncertainty and error in data, and a need to devise generic techniques to handle it that are appropriate within historical GIS.

Too often in the humanities, potentially useful technologies and methodologies have been developed, but have failed to make a real impact on mainstream scholarship (Boonstra *et al.*, 2004). There are many reasons for this, but at the core there is a cultural divide. On the one hand there are people that devise the technologies and methodologies, and, on the other, people who have the substantive research questions. There is a failure to cross this divide and see how the technologies and methodologies can be applied to advance the scholarship. Resolving this will require working in partnership, where the GIS experts focus their attention on areas that are of real need to historians, and historians are prepared to use and adopt appropriate data, technologies and methodologies. Collaborative working should be a major part of this. Humanities research has traditionally been a solitary activity; however, there are signs that this is changing. In the UK, the Arts and Humanities Research Council (AHRC), the main source of funding to the humanities, is actively encouraging team-based research. It does this through funding. In 2004–05, around three-quarters of its responsive mode funding went into collaborative research. The AHRC argues that one advantage of team-based research is that it allows technologies to be exploited (AHRC, 2005). This is certainly the case with historical GIS. In the US, there are more obstacles to collaborative work, as a result of the promotion and tenure systems regarding it with suspicion. This undoubtedly presents an obstacle to historical GIS, and it is hoped that attitudes will change in the near future.

There remains a more fundamental problem, and that is to identify what the technologies and methodologies have to offer to scholarship within the discipline. In the case of historical GIS, the answer is perhaps that it can stress the importance of geography, which has traditionally been neglected due to the difficulties of handling Geographical Information within source material. Although this answer appears straightforward, it poses the often asked question of 'why does geography matter?' or to take this further, 'why does the spatial data-driven version of geography offered by GIS matter?' If geography is taken to be the study of places, then GIS matters because it allows places to be subdivided into smaller components. The size of these components will vary according to the scale of the study area and the sources used. A national-level study can subdivide the country into districts or counties, and use census data to study these. At the opposite extreme, a local-level study area may subdivide into individuals and households using manuscript sources. Either way, the GIS allows the researcher to explore the study area not as an un-subdivided whole, but as a place that is made up by a large number of smaller places, each of which is affected by a slightly different process in the physical and human environment that surrounds it. The key challenge in using GIS is therefore to tell the story of the similarities and differences between the different places that make up the study area. If time is included in the study, this then becomes a challenge of telling the story of how different places developed differently, or responded differently to a certain phenomenon.

Perhaps surprisingly, this approach often leads to a conclusion that places were not as different as prevailing orthodoxies might suggest. Cunfer's work challenges the story that the Dust Bowl was caused by over-intensive agriculture, by showing that dust storms occurred in places that had little or no agriculture. Hillier showed that areas that had been redlined were not necessarily condemned to decline in the way that the orthodoxy suggests, by comparing these areas to areas that had not been redlined. Ray suggests that Salem was not as socially and economically divided as had been suggested. Many pieces of GIS research are likely to give similar outcomes. By subdividing study areas, they are able to explore whether the patterns that would have been expected to develop according to the common orthodoxy, actually did. It is thus an approach that allows convention to be challenged.

The question that follows from this is: 'if GIS can challenge accepted explanations, can it offer explanations of its own?' Here, the evidence is still limited. The answer may well depend on the source data used and on where the limits of what is, and is not, considered GIS-based research are drawn. Census-types analyses that rely on highly aggregate statistics are unlikely to offer new explanations through the use of, for example, multi-variate statistics. Instead, a more profitable avenue for explanation may well be to use these data to inform more detailed studies that may or may not use GIS. For example, a GIS analysis of census data may tell us which areas behaved in a certain way, and which seemed to behave differently. This can then be used to inform our use of local case studies, either to re-interpret existing studies or to choose sites that are worthy of further exploration. This further research is likely to use individual-level data, quite possibly without the use of GIS. With more detailed data, particularly if it is qualitative, this split will probably be less pronounced. Instead, it will be an intuitive part of the research process that certain places, perhaps down to the individual household, will be deemed to be important enough to study in detail, while others will be less important because they follow the same broad patterns. Again, the GIS will be a fundamental part of deciding which places to study in detail, and determining what their broader significance is.

Throughout this book, we have painted a picture of the use of a remarkable set of technologies and associated methodologies just as they begin to impact upon research debates in historical geography. To a degree, much of the hard work has already been done. GIS software is now considerably easier to use than it was in the early 1990s, when most software was command-line driven. Many key strategic datasets are now available to the historical geographer facilitating research that concentrates on exploring data and addressing historiographical debates, rather than collecting information. These key sources are complex and expensive resources, with many years of research time devoted to producing them and making them fit for use by researchers. Scholars now have the opportunity to use them. The remaining challenges are two-fold: firstly, to develop methodologies that make best use of the attribute, spatial and, where appropriate, temporal components of the available sources so as to turn these into usable information about the past; secondly, and more importantly, to take the technologies and methodologies offered by GIS and use these to reinvigorate the discipline of historical geography. This will only be done by delivering high-quality scholarship that significantly enhances our understanding of the key research debates. This is within our reach. We believe that if the opportunities currently on offer are taken, then GIS will become an essential part of historical research in the future.

REFERENCES

Abler R. A. (1987) 'What Shall We Say? To Whom Shall We Speak?' *Annals of the Association of American Geographers*, 77: pp. 511–524.

Abler R. A. (1993) 'Everything in its Place: GPS, GIS, and Geography in the 1990s' *Professional Geographer*, 45: pp. 131–139.

AHRC (2005) *Delivery Plan* (www.ahrc.ac.uk/about/delivery_plan.asp). Viewed 27 June 2007.

Al-Taha K. K., Snodgrass R. T. and Soo M. D. (1994) 'Bibliography on Spatio-Temporal Databases' *International Journal of Geographical Information Systems*, 8: pp. 95–103.

Andrienko G. and Andrienko N. (2001) 'Exploring Spatial Data with Dominant Attribute Map and Parallel Coordinates' *Computers Environment and Urban Systems*, 25: pp. 5–15.

Anon. (1990) 'ARC/INFO: An Example of a Contemporary Geographic Information System' in Peuquet D. J. and Marble D. F. (eds.) *Introductory Readings in Geographic Information Systems* Taylor & Francis: London, pp. 90–99.

Anselin L. (1995) 'Local Indicators of Spatial Association – LISA' *Geographical Analysis*, 27: pp. 93–115.

Anselin L., Syabri I. and Kho Y. (2006) 'GeoDa: An Introduction to Spatial Data Analysis' *Geographical Analysis*, 38: pp. 5–27.

Atkins P. and Mitchell G. (1996) 'Yesterday's and Today's Data' *GIS Europe*, 5: pp. 24–26.

Bailey T. C. and Gatrell A. C. (1995) *Interactive Spatial Data Analysis* Longman: Harlow.

Baker A. R. H. (2003) *Geography and History: Bridging the Divide* Cambridge University Press: Cambridge.

Bartley K. and Campbell B. (1997) 'Inquisitiones Post Mortem, GIS, and the Creation of a Land-use Map of Medieval England' *Transactions in GIS*, 2: pp. 333–346.

Bartley K., Ell P. S. and Lee J. (2000) 'From Manuscript to Multimedia' in *Data Modelling, Modelling History* Moscow University Press: Moscow.

Basoglu U. and Morrison J. L. (1978) 'An Efficient Hierarchical Data Structure for the US Historical Boundary File' in Dutton G. (ed.) *First International Advanced Study Symposium on Topological Data Structures for Geographic Information Systems. Volume IV – Data Structures: Hierarchical and Overlaid* Laboratory for Computer Graphics and Spatial Analysis, Harvard University: Cambridge, MA.

Berman F. (2003) *Grid Computing: Making the Global Infrastructure a Reality* John Wiley: Chichester.

Berman M. L. (2005) 'Boundaries or Networks in Historical GIS: Concepts of Measuring Space and Administrative Geography in Chinese History' *Historical Geography*, 33: pp. 118–133.

Bernhardsen T. (1999) 'Choosing a GIS' in Longley P. A., Goodchild M. F., Maguire D. J. and Rhind D. W. (eds.) *Geographical Information Systems: Principals, Techniques, Management and Applications*, 2nd edition John Wiley: Chichester, pp. 589–600.

Beveridge A. A. (2002) 'Immigration, Ethnicity and Race in Metropolitan New York, 1900–2000' in Knowles A. K. (ed.) *Past time, Past Place: GIS for History* ESRI Press: Redlands, CA, pp. 65–78.

Block W. and Thomas W. (2003) 'Implementing the Data Documentation Initiative at the Minnesota Population Center' *Historical Methods*, 36: pp. 97–101.

Bol P. and Ge J. (2005) 'China Historical GIS' *Historical Geography*, 33: pp. 150–152.

Bondi L. and Bradford M. (1990) 'Applications of Multi-level Modelling to Geography' *Area*, 22: pp. 256–263.

Boonstra O., Breure L. and Doorn P. (2004) *Past, Present and Future of Historical Information Science* NIWI-KNAW: Amsterdam.

Boyer P. S. and Nissenbaum S. (1974) *Salem Possessed: the Social Origins of Witchcraft* Harvard University Press: Cambridge, MA.

Bracken I. (1994) 'A Surface Model Approach to the Representation of Population-related Social Indicators' in Fotheringham A. S. and Rogerson P. (eds.) *Spatial Analysis and GIS* Taylor & Francis: London, pp. 247–260.

Bracken I. and Martin D. (1989) 'The Generation of Spatial Population Distributions from Census Centroid Data' *Environment and Planning A*, 21: pp. 537–453.

Bracken I. and Martin D. (1995) 'Linkage of the 1981 and 1991 UK Census Using Surface Modelling Concepts' *Environment and Planning A*, 27: pp. 379–390.

Brewer C. (2002) *ColorBrewer: A Web Tool for Selecting Colors for Maps* (www.colorbrewer.org). Viewed 22 June 2007.

Brimicombe A. (1998) 'A Fuzzy Coordinate System for Locational Uncertainty in Space and Time' in Carver S. (ed.) *Innovations in GIS 5* Taylor & Francis: London, pp. 141–150.

Brunsdon C., Fotheringham A. S. and Charlton M. E. (1996) 'Geographically Weighted Regression: a Method for Exploring Spatial Nonstationarity' *Geographical Analysis*, 28: pp. 281–298.

Buckatzsch E. (1950) 'The Geographical Distribution of Wealth in England, 1086–1843: an Experimental Study of Certain Tax Assessments' *Economic History Review: Second Series*, 3: pp. 180–202.

Buckland M. and Lancaster L. (2004) 'Combining Place, Time, and Topic' *D-Lib Magazine*, 10 (5) (www.dlib.org). Viewed 22 June 2007.

Bugayevskiy L. M. and Snyder J. P. (1995) *Map Projections: a Reference Manual* Taylor & Francis: London.

Burrough P. A. and McDonnell R. (1998) *Principals of Geographical Information Systems for Land Resources Assessment*, 2nd edition Clarendon Press: Oxford.

Butlin R. A. (1993) *Historical Geography: through the Gates of Space and Time* Edward Arnold: London.

Camara A. S. and Raper J. (1999, eds.) *Spatial Multimedia and Virtual Reality* Taylor & Francis: London.

Campbell B. M. S. (2000) *English Seigniorial Agriculture 1250–1450* Cambridge University Press: Cambridge.

Campbell B. M. S. and Bartley K. (2006) *England on the Eve of the Black Death: An Atlas of Lay Lordship, Land and Wealth, 1300–1349* Manchester University Press: Manchester.

Campos R. M., Congdon P., Curtis S. E., Gregory I. N., Jones I. R. and Southall H. R. (2004) 'Locality Level Mortality and Socio-economic Change in Britain since 1920: First Steps towards Analysis of Infant Mortality Variation' in Boyle P., Curtis S. E., Graham E., and Moore E. (eds.) *The Geography of Health Inequalities in the Developed World* Ashgate: Aldershot, pp. 53–75.

Carver S. and Openshaw S. (1992) 'A Geographic Information Systems Approach to Locating Nuclear Waste Disposal Sites' in Clark M., Smith D. and Blowers A. (eds.) *Waste Location: Spatial Aspects of Waste Management, Hazards and Disposal* Routledge: London, pp. 105–127.

Chorley R. and Buxton R. (1991) 'The Government Setting of GIS in the United Kingdom' in Maguire D. J., Goodchild M. F. and Rhind D. W. (eds.) *Geographical Information Systems: Principles and Applications. Volume I: Principles* Longman Scientific and Technical: London, pp. 67–79.

Chrisman N. R. (1990) 'The Accuracy of Map Overlays: a Reassessment' in Peuquet D. J. and Marble D. F. (eds.) *Introductory Readings in Geographic Information Systems* Taylor & Francis: London, pp. 308–320.

Chrisman N. R. (1999) 'What does "GIS" Mean?' *Transactions in GIS*, 3: pp. 175–186.

Chrisman N. R. (2002) *Exploring Geographic Information Systems*, 2nd edition John Wiley: Chichester.

Clarke K. C. (1997) *Getting Started with Geographic Information Systems* Prentice Hall: Upper Saddle River, NJ.

Cliff A. D. and Ord J. K. (1973) *Spatial Autocorrelation* Pion: London.

Cockings S., Fisher P. F., and Langford M. (1997) 'Parameterization and Visualization of the Errors in Areal Interpolation' *Geographical Analysis*, 29: pp. 314–328.

Congdon P., Campos R. M., Curtis S. E., Southall H. R., Gregory I. N. and Jones I. R. (2001) 'Quantifying and Explaining Changes in Geographical Inequality of Infant Mortality in England and Wales since the 1890s' *International Journal of Population Geography*, 7: pp. 35–51.

Coppock J. T. and Rhind D. W. (1991) 'The History of GIS' in Maguire D. J., Goodchild M. F. and Rhind D. W. (eds.) *Geographical Information Systems: Principles and Applications. Volume I: Principles* Longman Scientific and Technical: London, pp. 21–43.

Couclelis H. (1998) 'Geocomputation in Context' in Longley P. A., Brooks S. M., McDonnell R. and MacMillan B. (eds.) *Geocomputation: a Primer* John Wiley: London, pp. 17–29.

Cowen D. (1990) 'GIS versus CAD versus DBMS: What are the Differences?' in Peuquet D. J. and Marble D. F. (eds.) *Introductory Readings in Geographical Information Systems* Taylor & Francis: London, pp. 52–62.

Cox N. J. and Jones K. (1981) 'Exploratory Data Analysis' in Wrigley N. and Bennett R. J. (eds.) *Quantitative Geography* Edward Arnold: London, pp. 135–143.

Cresswell T. (2004) *Place: a Short Introduction* Blackwell Publishing: Oxford.

Cunfer G. (2002) 'Causes of the Dust Bowl' in Knowles A. K. (ed.) *Past Time, Past Place: GIS for History* ESRI Press: Redlands, CA, pp. 93–104.

Cunfer G. (2004) 'Dust Storms before the Dust Bowl' Paper presented to the twenty-ninth Annual Meeting of the Social Science History Association.

Cunfer G. (2005) *On the Great Plains: Agriculture and Environment* Texas A&M University Press: College Station, TX.

Curry M. (1995) 'GIS and the Inevitable Ethical Inconsistency' in Pickles J. (ed.) *Ground Truth: the Social Implications of Geographic Information Systems* Guilford Press: New York, NY, pp. 68–87.

Darby H. C. (1977) *Domesday England* Cambridge University Press: Cambridge.

Darby H. C. and Versey G. R. (1975) *Domesday Gazetteer* Cambridge University Press: Cambridge.

Darby H. C., Glasscock R., Sheail J. and Versey G. R. (1979) 'The Changing Geographical Distribution of Wealth in England: 1086–1334–1525' *Journal of Historical Geography*, 5: pp. 247–262.

Date C. J. (1995) *An Introduction to Database Systems*, 6th edition Addison-Wesley: Reading, MA.

DDI (2006) *Data Documentation Initiative* (www.icpsr.umich.edu/DDR). Viewed 22 June 2007.

De Moor M. and Wiedemann T. (2001) 'Reconstructing Belgian Territorial Units and Hierarchies: an Example from Belgium' *History and Computing*, 13: pp. 71–97.

DeMers M. N. (2002) *GIS Modeling in Raster* John Wiley: Chichester.

Department of the Environment (1987) *Handling Geographic Information: Report of the Committee of Enquiry chaired by Lord Chorley* HMSO: London.

Diamond E. and Bodenhamer D. (2001) 'Investigating White-flight in Indianapolis: a GIS Approach' *History and Computing*, 13: pp. 25–44.

Dodgshon R. A. (1998) *Society in Time and Space: a Geographical Perspective on Change* Cambridge University Press: Cambridge.

Dorling D. (1992) 'Visualising People in Space and Time' *Environment and Planning B*, 19: pp. 613–637.

Dorling D. (1995) *A New Social Atlas of Britain* John Wiley & Sons: Chichester.

Dorling D. (1996) 'Area Cartograms: their Use and Creation' *Concepts and Techniques in Modern Geography*, 59. University of East Anglia, Environmental Publications: Norwich.

Dorling D. and Fairburn D. (1997) *Mapping: Ways of Representing the World* Longman: Harlow.

Dublin Core Metadata Initiative (2005) *Dublin Core Metadata Initiative* (http://dublincore.org). Viewed 22 June 2007.

Earickson R. and Harlin J. (1994) *Geographic Measurement and Quantitative Analysis* MacMillan: New York, NY.

Ebdon D. (1985) *Statistics in Geography*, 2nd edition Basil Blackwell: Oxford.

ECAI (2005) *The Electronic Cultural Atlas Initiative* (www.ecai.org). Viewed 22 June 2007.

Egenhofer M. J. and Herring J. R. (1991) 'High-level Spatial Data Structures for GIS' in Maguire D. J., Goodchild M. F. and Rhind D. W. (eds.) *Geographical Information Systems: Principles and Applications. Volume I: Principles* Longman Scientific and Technical: Harlow, pp. 227–237.

Ell P. S. and Gregory I. N. (2001) 'Adding a New Dimension to Historical Research with GIS' *History and Computing*, 13: pp. 1–6.

ESRI (1994) *ArcInfo Data Management: Concepts, Data Models, Database Design, and Storage* Environmental Systems Research Institute: Redlands, CA.

ESRI (2005) *ESRI GIS and Mapping Software* (www.esri.com). Viewed 22 June 2007.

Evans, I. S. (1977) 'The Selection of Class Intervals' *Transactions of the Institute of British Geographers*, 2: pp. 98–124.

FGDC (2005) *The Federal Geographic Data Committee* (www.fgdc.gov). Viewed 22 June 2007.

Fischer M. M., Scholten H. J. and Unwin D. (1996) 'Geographic Information Systems, Spatial Data Analysis and Spatial Modelling: an Introduction' in Fischer M. M., Scholten H. J. and Unwin D. (eds.) *Spatial Analytical Perspectives on GIS* Taylor & Francis: London, pp. 3–20.

Fisher P. F. (1999) 'Models of Uncertainty in Spatial Data' in Longley P. A., Goodchild M. F., Maguire D. J. and Rhind D. W. (eds.) *Geographical Information Systems: Principals, Techniques, Management and Applications*, 2nd edition John Wiley: Chichester pp. 191–205.

Fisher P. F. and Unwin D. (2002) *Virtual Reality in Geography* Taylor & Francis: London.

Fitch C. A. and Ruggles S. (2003) 'Building the National Historical Geographic Information System' *Historical Methods*, 36: pp. 41–51.

Fleming R. and Lowerre A. (2004) 'MacDomesday Book' *Past and Present*, 184: pp. 209–232.

Flowerdew R. F. (1991) 'Spatial Data Integration' in Maguire D. J., Goodchild M. F. and Rhind D. W. (eds.) *Geographical Information Systems: Principles and Applications. Volume I: Principles* Longman Scientific and Technical: Harlow, pp. 375–387.

Flowerdew R. and Green M. (1994) 'Areal Interpolation and Types of Data' in Fotheringham A. S. and Rogerson P. A. (eds.) *Spatial Analysis and GIS* Taylor & Francis: London, pp. 121–145.

Flowerdew R., Geddes A. and Green M. (2001) 'Behaviour of Regression Models under Random Aggregation' in Tate N. J. and Atkinson P. M. (eds.) *Modelling Scale in Geographical Information Science* John Wiley & Sons: Chichester, pp. 89–104.

Fotheringham A. S. (1997) 'Trends in Quantitative Methods I: Stressing the Local' *Progress in Human Geography*, 21: pp. 88–96.

Fotheringham A. S. and Wong D. (1991) 'The Modifiable Areal Unit Problem in Multi-variant Statistical Analysis' *Environment and Planning A*, 23: pp. 1025–1044.

Fotheringham A. S., Brunsdon C. and Charlton M. E. (2000) *Quantitative Geography: Perspectives on Spatial Data Analysis* Sage: London.

Fotheringham A. S., Brunsdon C. and Charlton M. E. (2002) *Geographically Weighted Regression: the Analysis of Spatially Varying Relationships* John Wiley & Sons: Chichester.

Fowler M. (1996) 'High-resolution Satellite Imagery in Archaeological Application – a Russian Satellite Photograph of the Stonehenge Region' *Antiquity*, 70: pp. 667–671.

Friedlander D. and Roshier R. (1965) 'A Study of Internal Migration in England and Wales, Part 1: Geographical Patterns of Internal Migration, 1851–1951' *Population Studies*, 19: pp. 239–280.

Gastner M. T. and Newman M. E. J. (2004) 'Diffusion-based Method for Producing Density Equalising Maps' *Proceedings of the National Academy of Sciences*, 101: pp. 7499–7503.

Gastner M. T., Shalizi C. R. and Newman M. E. J. (2004) *Maps and Cartograms of the 2004 Election Results* (www-personal.umich.edu/~mejn/election). Viewed 22 June 2007.

Gatley D. A. and Ell P. S. (2000) *Counting Heads: an Introduction to the Census, Poor Law Union Data and Vital Registration* Statistics for Education: York.

Gatrell A. C. (1983) *Distance and Space: a Geographical Perspective* Clarendon Press: Oxford.

Gatrell A. C. (1985) 'Any Space for Spatial Analysis?' in Johnson R. J. (ed.) *The Future of Geography* Methuen: London, pp. 190–208.

Gatrell A. C. (1991) 'Concepts of Space and Geographical Data' in Maguire D. J., Goodchild, M. F. and Rhind D. W. (eds.) *Geographical Information Systems: Principles and Applications. Volume I: Principles* Longman Scientific and Technical: Harlow, pp. 119–134.

Geography Network (2005) *The Geography Network* (www.geographynetwork.com). Viewed 22 June 2007.

Getis A. and Ord J. K. (1992) 'The Analysis of Spatial Association by Use of Distance Statistics' *Geographical Analysis*, 24: pp. 189–206.

Getis A. and Ord J. K. (1996) 'Local Spatial Statistics: an Overview' in Longley P. and Batty M. (eds.) *Spatial Analysis: Modelling in a GIS Environment* GeoInformation International: Cambridge, pp. 261–277.

Gilbert D. M. and Southall H. R. (1996) 'Indicators of Regional Economic Disparity: the Geography of Economic Distress in Britain before 1914' in Harvey C. and Press J. (eds.) *Databases in Historical Geography* MacMillan: Basingstoke, pp. 140–146.

Goerke M. (1994, ed.) *Coordinates for Historical Maps* Max-Planck-Institut fur Geschichte: Gottingen.

Goodchild M. F. (1987) 'Introduction to Spatial Autocorrelation' *Concepts and Techniques in Modern Geography*, 47. GeoAbstracts: Norwich.

Goodchild M. F. (1992a) 'Geographical Information Science' *International Journal of Geographical Information Systems*, 6: pp. 31–45.

Goodchild M. F. (1992b) *Research Initiative 1: Accuracy of Spatial Databases. Final Report* NCGIA, University of California: Santa Barbara, CA.

Goodchild M. F. (2004) 'The Alexandria Digital Library Project: Review, Assessment, and Prospects' *D-Lib Magazine*, 10 (5): (www.dlib.org). Viewed 22 June 2007.

Goodchild M. F. and Gopal S. (1989, eds.) *The Accuracy of Spatial Databases* Taylor & Francis: London.

Goodchild M. F. and Lam N. S.-N. (1980) 'Areal Interpolation: a Variant of the Traditional Spatial Problem' *Geo-Processing*, 1: pp. 297–312.

Goodchild M. F., Anselin L. and Deichmann U. (1993) 'A Framework for the Areal Interpolation of Socio-economic Data' *Environment and Planning A*, 25: pp. 383–397.

Goodchild M. F., Chi-Chang L. and Leung Y. (1994) 'Visualising Fuzzy Maps' in Hearnshaw H. M. and Unwin D. J. (eds.) *Visualisation in Geographical Information Systems* John Wiley: Chichester, pp. 158–167.

Goovaerts P. (2002) 'Geostatistical Approaches to Spatial Uncertainty Using *p*-field Simulation with Conditional Probability Fields' *International Journal of Geographical Information Science*, 16: 167–178.

Green D. and Bossomaier T. (2002) *Online GIS and Spatial Metadata* Taylor & Francis: London.

Gregory I. N. (2000) 'Longitudinal Analysis of Age and Gender Specific Migration Patterns in England and Wales: a GIS-based Approach' *Social Science History*, 24: pp. 471–503.

Gregory I. N. (2002a) 'Time Variant Databases of Changing Historical Administrative Boundaries: a European Comparison' *Transactions in GIS*, 6: pp. 161–178.

Gregory I. N. (2002b) 'The Accuracy of Areal Interpolation Techniques: Standardising 19th and 20th century Census Data to Allow Long-term Comparisons' *Computers Environment and Urban Systems*, 26: pp. 293–314.

Gregory I. N. (2003) *A Place in History: a Guide to Using GIS in Historical Research* Oxbow: Oxford.

Gregory I. N. (2005) 'The Great Britain Historical GIS' *Historical Geography*, 33: pp. 132–134.

Gregory I. N. (2007, in press) '"A Map is Just a Bad Graph": Spatial Analysis of Quantitative Data in Historical GIS" in Knowles A. K. *Placing History: How Maps, Spatial Data and GIS are Changing Historical Scholarship.*

Gregory I. N. and Ell P. S. (2005a) 'Breaking the Boundaries: Integrating 200 years of the Census using GIS' *Journal of the Royal Statistical Society, Series A*, 168: pp. 419–437.

Gregory I. N. and Ell P. S. (2005b) 'Analysing Spatio-temporal Change Using National Historical GISs: Population Change during and after the Great Irish Famine' *Historical Methods*, 38: pp. 149–167.

Gregory I. N. and Ell P. S. (2006) 'Error Sensitive Historical GIS: Identifying Areal Interpolation Errors in Time-series Data' *International Journal of Geographical Information Science*, 20: pp. 135–152.

Gregory I. N. and Southall H. R. (1998) 'Putting the Past in its Place: the Great Britain Historical GIS' in Carver S. (ed.) *Innovations in GIS 5* Taylor & Francis: London: pp. 210–221.

Gregory I. N. and Southall H. R. (2000) 'Spatial Frameworks for Historical Censuses – the Great Britain Historical GIS' in Hall P. K., McCaa R. and Thorvaldsen D. (eds.) *Handbook of Historical Microdata for Population Research* Minnesota Population Center: Minneapolis, MN, pp. 319–333.

Gregory I. N., Bennett C., Gilham V. L. and Southall H. R. (2002) 'The Great Britain Historical GIS: from Maps to Changing Human Geography' *The Cartographic Journal*, 39: pp. 37–49.

Gregory I. N., Kemp K. K. and Mostern R. (2001a) 'Geographic Information and Historical Research: Current Progress and Future Directions' *History and Computing*, 13: pp. 7–21.

Gregory I. N., Dorling D. and Southall H. R. (2001b) 'A Century of Inequality in England and Wales using Standardised Geographical Units' *Area*, 33: pp. 297–311.

Guelke L. (1997) 'The Relations between Geography and History Reconsidered' *History and Theory*, 36: pp. 216–234.

Guptil S. C. (1999) 'Metadata and Data Catalogues' in Longley P. A., Goodchild M. F., Maguire D. J. and Rhind D. W. (eds.) *Geographical Information Systems. Volume I: Principles and Technical Issues* John Wiley: Chichester, pp. 677–692.

Hall S. S. (1992) *Mapping the Next Millennium: how Computer-driven Cartography is Revolutionizing the Face of Science* Vintage Books: New York, NY.

Harder C. (1998) *Serving Maps on the Internet: Geographic Information on the World Wide Web* ESRI Press: Redlands, CA.

Harpring P. (1997) 'Proper Words in Proper Places: the Thesaurus of Geographic Names' *MDA Information*, 2: pp. 5–12.

Harris R. (1987, ed.) *The Historical Atlas of Canada. Volume I: from the Beginning to 1800* University of Toronto Press: Toronto, ON.

Harris T. (2002) 'GIS in Archaeology' in Knowles A. K. (ed.) *Past Time, Past Place: GIS for History* ESRI Press: Redlands, CA, pp. 131–143.

Harris T., Turley B. and Rouse L. J. (2005) *Integrating the Humanities and Geospatial Science: Exploring Cultural Resources and Sacred Space through Internet GIS* (http://ark.geo.wvu.edu/projects.html). Viewed 22 June 2007.

Harley J. B. (1989) 'Deconstructing the Map' *Cartographica*, 26: pp. 11–20.

Harley J. B. (1990) 'Cartography, Ethics and Social Theory' *Cartographica*, 27: pp. 1–23.

Harvey C. and Press J. (1996, eds.) *Databases in Historical Research: Theory, Methods and Applications* MacMillan Press: Basingstoke.

Healey R. G. (1991) 'Database Management Systems' in Maguire D. J., Goodchild, M. F. and Rhind D. W. (eds.) *Geographical Information Systems: Principles and Applications. Volume I: Principles* Longman Scientific and Technical: Harlow, pp. 251–267.

Healey R. G. and Stamp T. R. (2000) 'Historical GIS as a Foundation for the Analysis of Regional Economic Growth: Theoretical, Methodological, and Practical Issues' *Social Science History*, 24: pp. 575–612.

Hernandez M. J. (1997) *Database Design for Mere Mortals* Addison-Wesley: Reading, MA.

Heuvelink G. B. M. (1999) 'Propagation of Error in Spatial Modelling with GIS' in Longley P. A., Goodchild M. F., Maguire D. J. and Rhind D. W. (eds.) *Geographical Information Systems: Principals, Techniques, Management and Applications*, 2nd edition John Wiley: Chichester pp. 207–217.

Heuvelink G. B. M., Burrough P. A. and Stein A. (1989) 'Propagation of Errors in Spatial Modelling with GIS' *International Journal of Geographical Information Systems*, 3: pp. 303–322.

Heywood D. I., Cornelius S. and Carver S. (2002) *An Introduction to Geographic Information Systems*, 2nd edition Longman: Harlow.

Hill L. L. (2004) 'Guest Editorial: Georeferencing in Digital Libraries' *D-Lib Magazine*, 10 (5) (www.dlib.org). Viewed 22 June 2007.

Hill L. L. and Janee G. (2004) 'The Alexandria Digital Library Project: Metadata Development and Use' in Hillmann D. I. and Westbrooks E. L. (eds.) *Metadata in Practice* American Library Association: Chicago, pp. 117–138.

Hillier A. E. (2002) 'Redlining in Philadelphia' in Knowles, A. K. (ed.) *Past Time, Past Place: GIS for History* ESRI: Redlands, CA, pp. 79–93.

Hillier A. E. (2003) 'Spatial Analysis of Historical Redlining: a Methodological Exploration' *Journal of Housing Research*, 14: pp. 137–167.

Hoffman-Wellenhof B. (1997) *Global Positioning System: Theory and Practice*, 4th edition Springer Verlag: Vienna.

Holdsworth D. W. (2002) 'Historical Geography: the Ancients and the Moderns – Generational Vitality' *Progress in Human Geography*, 26: pp. 671–678.

Holdsworth D. W. (2003) 'Historical Geography: New Ways of Imaging and Seeing the Past' *Progress in Human Geography*, 27: pp. 486–493.

IDP (2005) *The International Dunhuang Project* (http://idp.bl.uk). Viewed 22 June 2007.

Intra-Governmental Group on Geographic Information (2004) *The Principles of Good Metadata Management* Office of the Deputy Prime Minister: London.

Isaaks E. H. and Srivastava R. M. (1989) *An Introduction to Applied Geostatistics* Oxford University Press: Oxford.

Jenks G. F. and Caspall F. C. (1971) 'Error on Choroplethic Maps: Definition, Measurement, Reduction' *Annals of the Association of American Geographers*, 61: pp. 217–243.

Jensen J. R. (1996) *Introductory Digital Image Processing: a Remote Sensing Perspective*, 2nd edition Prentice Hall: Upper Saddle River, NJ.

Johnston R. J. (1983) *Philosophy and Human Geography: an Introduction to Contemporary Approaches* Edward Arnold: London.

Johnston R. J. (1999) 'Geography and GIS' in Longley P. A., Goodchild M. F., Maguire D. J. and Rhind D. W. (eds.) *Geographical Information Systems: Principals, Techniques, Management and Applications*, 2nd edition Chichester: John Wiley, pp. 39–47.

Jones K. (1991a) 'Specifying and Estimating Multilevel Models for Geographical Research' *Transactions of the Institute of British Geographers*, 16: pp. 148–159.

Jones K. (1991b) 'Multilevel Models for Geographical Research' *Concepts and Techniques in Modern Geography*, 54. Environmental Publications: Norwich.

Jones C. (1997) *Geographic Information Systems and Computer Cartography* Longman: Harlow.

Jones R. (2004) 'What Time Human Geography?' *Progress in Human Geography*, 28: pp. 287–304.

Jones C. B., Alani H. and Tudhope D. (2003) 'Geographical Terminology Servers – Closing the Semantic Divide' in Duckham M., Goodchild M. F. and Worboys M. F. (eds.) *Foundations of Geographic Information Science* Taylor & Francis: London, pp. 205–222.

Kain R. J. P. and Prince H. C. (2000) *Tithe Surveys for Historians* Phillimore: Chichester.

Kampke T. (1994) 'Storing and Retrieving Changes in a Sequence of Polygons' *International Journal of Geographic Information Systems*, 8: pp. 493–513.

Keates J. (1989) *Cartographic Design and Production,* 2nd edition Longman Scientific and Technical: Harlow.

Kennedy L., Ell P. S., Crawford E. M. and Clarkson L. A. (1999) *Mapping the Great Irish Famine: an Atlas of the Famine Years* Four Courts Press: Dublin.

Kennedy M. (2002) *The Global Positioning System and GIS,* 2nd edition Taylor & Francis: London.

Kerr D. and Holdsworth D. (1990, eds.) *The Historical Atlas of Canada. Volume III: Addressing the Twentieth Century* University of Toronto Press: Toronto, ON.

Knowles A. K. (2000) 'Introduction' *Social Science History,* 24: pp. 451–470.

Knowles A. K. (2002a) 'Introducing Historical GIS' in Knowles A. K. (ed.) *Past Time, Past Place: GIS for History* ESRI Press: Redlands, CA, pp. xi–xx.

Knowles A. K. (2002b, ed.) *Past Time, Past Place: GIS for History* ESRI Press: Redlands, CA.

Knowles A. K. (2004) 'Visualizing Gettysburg: Problems in Historical Terrain Modeling' Paper presented at the 2004 Social Science History Association conference.

Knowles A. K. (2005a) 'Emerging Trends in Historical GIS' *Historical Geography,* 33: pp. 7–13.

Knowles A. K. (2005b, ed.) 'Reports on National Historical GIS Projects' *Historical Geography,* 33: pp. 293–314.

Kollias V. J. and Voliotis A. (1991) 'Fuzzy Reasoning in the Development of Geographical Information Systems' *International Journal of Geographical Information Systems,* 5: pp. 209–223.

Krygier J. B. (1997) 'Envisioning the American West: Maps, the Representational Barrage of 19th century Expedition Reports, and the Production of Scientific Knowledge' *Cartography and Geographic Information Systems,* 24: pp. 27–50.

Lancaster L. and Bodenhamer D. (2002) 'The Electronic Cultural Atlas Initiative and the North American Religion Atlas' in Knowles A. K. (ed.) *Past time, Past Place: GIS for History* ESRI Press: Redlands, CA, pp. 163–178.

Langford M., Maguire D. and Unwin D. J. (1991) 'The Areal Interpolation Problem: Estimating Population using Remote Sensing in a GIS Framework' in Masser I. and Blakemore M. (eds.) *Handling Geographical Information: Methodology and Potential Applications* Longman: Harlow, pp. 55–77.

Langran G. (1992) *Time in Geographic Information Systems* Taylor & Francis: London.

Langran G. and Chrisman N. (1988) 'A Framework for Temporal Geographic Information' *Cartographica,* 25: pp. 1–14.

Langton J. (1972) 'Systems Approach to Change in Human Geography' *Progress in Geography,* 4: pp. 123–178.

Larson R. R. (2003) 'Placing Cultural Events and Documents in Space and Time' in Duckham M., Goodchild M. F. and Worboys M. F. (eds.) *Foundations of Geographic Information Science* Taylor & Francis: London, pp. 223–239.

Lawton R. (1968) 'Population Changes in England and Wales in the Later Nineteenth Century: an Analysis of Trends by Registration District' *Transactions of the Institute of British Geographers,* 44: pp. 55–74.

Lawton R. (1970) 'The Population of Liverpool in the Late Nineteenth Century' in Baker A., Hamshere J. and Langton J. (eds.) *Geographical Interpretations of Historical Sources* David & Charles: Newton Abbot, pp. 381–415.

Lee C. H. (1979) *British Regional Employment Statistics, 1841–1971* Cambridge University Press: Cambridge.

Lee C. H. (1991) 'Regional Inequalities in Infant Mortality in Britain, 1861–1971: Patterns and Hypotheses' *Population Studies*, 45: pp. 55–65.

Lillesand T. M. and Kiefer R. W. (2000) *Remote Sensing and Image Interpretation*, 4th edition John Wiley & Sons: Chichester.

Lilley K., Lloyd C., Trick S. and Graham C. (2005a) 'Mapping and Analyzing Medieval Built Form Using GPS and GIS' *Urban Morphology*, 9: 5–15.

Lilley K., Lloyd, C. and Trick S. (2005b) 'Mapping Medieval Urban Landscapes: the Design and Planning of Edward I's New Towns of England and Wales' *Antiquity*, 79 (303) (www.antiquity.ac.uk). Viewed 22 June 2007.

Lilley K., Lloyd, C. and Trick S. (2005c) *Mapping Medieval Townscapes: a Digital Atlas of the New Towns of Edward I* (http://ads.ahds.ac.uk/catalogue/specColl/atlas_ahrb_2005/index.cfm). Viewed 22 June 2007.

Long J. H. (1994) 'Atlas of Historical County Boundaries' *Journal of American History*, 81: pp. 1859–1863.

Longley P. A., Goodchild M. F., Maguire D. J. and Rhind D. W. (2001) *Geographical Information Systems and Science* John Wiley & Sons: Chichester.

Louis R. C. (1993, ed.) *The Historical Atlas of Canada. Volume II: the Land Transformed, 1800–1891* University of Toronto Press: Toronto, ON.

Lowe D. W. (2002) 'Telling Civil War Battlefield Stories with GIS' in Knowles A. K. (ed.) *Past Time, Past Place: GIS for History* ESRI Press: Redlands, CA, pp. 51–63.

MacDonald B. M. and Black F. A. (2000) 'Using GIS for Spatial and Temporal Analysis in Print Culture Studies: Some Opportunities and Challenges' *Social Science History*, 24: pp. 505–536.

MacEachren A. M. (1994) 'Time as a Cartographic Variable' in Hearnshaw H. M. and Unwin D. J. (eds.) *Visualisation in GIS* John Wiley & Sons: Chichester, pp. 115–130.

MacEachren A. M. (1995) *How Maps Work: Representation, Visualisation and Design* Guildford Press: London.

Madry S. L. H. and Crumley C. L. (1990) 'An Application of Remote Sensing and GIS in a Regional Archaeological Settlement Analysis: the Arroux River Valley, Burgundy, France' in Allen K. M. S., Green S. W. and Zubrow E. B. W. (eds.) *Interpreting Space: GIS and Archaeology* Taylor & Francis: London, pp. 364–381.

Martin D. (1996a) *Geographic Information Systems and their Socio-economic Applications*, 2nd edition Routledge: Hampshire.

Martin D. (1996b) 'Depicting Changing Distributions through Surface Estimations' in Longley P. and Batty M. (eds.) *Spatial Analysis: Modeling in a GIS environment* GeoInformation International: Cambridge, pp. 105–122.

Martin D. and Bracken I. (1991) 'Techniques for Representing Population-related Raster Databases' *Environment and Planning A*, 23: pp. 1069–1075.

Martin D., Dorling D. and Mitchell R. (2002) 'Linking Censuses through Time: Problems and Solutions' *Area*, 34: pp. 82–91.

Masser I. (1988) 'The Academic Setting of GIS' *International Journal of Geographical Information Systems*, 2: pp. 11–22.

Massey D. (1999) 'Space-time, "science" and the Relationship between Physical Geography and Human Geography' *Transactions of the Institute of British Geographers: New Series*, 24: pp. 261–276.

Massey D. (2005) *For Space* Sage: London.

McMaster R. B. and Noble P. (2005) 'The US National Historical Geographical Information System' *Historical Geography*, 33: pp. 134–136.

Miller D. and Modell J. (1988) 'Teaching United States History with the Great American History Machine' *Historical Methods*, 21: pp. 121–134.

Miller P. (1953) *The New England Mind: from Colony to Province* Harvard University Press: Cambridge, MA.

Miller P. and Greenstein D. (1997) *Discovering Online Resources across the Humanities: a Practical Implementation of the Dublin Core* UKOLN: Bath.

Monmonier M. S. (1996) *How to Lie with Maps*, 2nd edition University of Chicago Press: Chicago, IL.

Monmonier M. S. and Schnell G. A. (1988) *Map Appreciation* Prentice Hall: Upper Saddle River, NJ.

Morphet C. (1993) 'The Mapping of Small Area Data – a Consideration of the Role of Enumeration District boundaries' *Environment and Planning A*, 25: pp. 267–278.

Norman P., Rees P. and Boyle P. (2003) 'Achieving Data Compatibility over Space and Time: Creating Consistent Geographical Zones' *International Journal of Population Geography*, 9: pp. 365–386.

Ogborn M. (1999) 'The Relations between Geography and History: Work in Historical Geography in 1997' *Progress in Human Geography*, 23: pp. 97–108.

Openshaw S. (1984) 'The Modifiable Areal Unit Problem' *Concepts and Techniques in Modern Geography*, 38. Geobooks: Norwich.

Openshaw S. (1991a) 'A View on the GIS Crisis in Geography or Using GIS to Put Humpty-Dumpty Back Together Again' *Environment and Planning A*, 23: pp. 621–628.

Openshaw S. (1991b) 'Developing Appropriate Spatial Analysis Methods for GIS' in Maguire D. J., Goodchild M. F. and Rhind D. W. (eds.) *Geographical Information Systems: Principles and Applications. Volume I: Principles* Longman Scientific and Technical: London, pp. 389–401.

Openshaw S. (1992) 'Further Thoughts on Geography and GIS – a Reply' *Environment and Planning A*, 24: pp. 463–466.

Openshaw S. (1996) 'Developing GIS-relevant Zone-based Spatial Analysis Methods' in Longley P. and Batty M. (eds.) *Spatial Analysis: Modeling in a GIS Environment* GeoInformation International: Cambridge, pp. 55–74.

Openshaw S. (1997) 'The Truth about Ground Truth' *Transactions in GIS*, 2: pp. 2–24.

Openshaw S. (2000) 'GeoComputation' in Openshaw S. and Abrahart R. J. (eds.) *GeoComputation* Taylor & Francis: London, pp. 1–32.

Openshaw S. and Clarke G. (1996) 'Developing Spatial Analysis Functions Relevant to GIS Environments' in Fischer M., Scholten H. J. and Unwin D. (eds.) *Spatial Analytical Perspectives on GIS* Taylor & Francis: London, pp. 21–38.

Openshaw S. and Rao L. (1995) 'Algorithms for Re-aggregating 1991 Census Geography' *Environment and Planning A*, 27: pp. 425–446.

Openshaw S. and Taylor P. J. (1979) 'A Million or so Correlation Coefficients: Three Experiments on the Modifiable Areal Unit Problem' in Wrigley N. (ed.) *Statistical Applications in the Spatial Sciences* Pion: London, pp. 127–144.

Openshaw S., Charlton M., Wymer C. and Craft A. W. (1987) 'A Mark 1 Geographical Analysis Machine for the Automated Analysis of Point Data Sets' *International Journal of Geographical Information Systems*, 1: pp. 335–358.

Openshaw S., Waugh D. and Cross A. (1994) 'Some Ideas about the Use of Map Animation as a Spatial Analysis Tool' in Hearnshaw H. M. and Unwin D. J. (eds.) *Visualisation in GIS* John Wiley & Sons: Chichester, pp. 131–138.

Oracle (2005) *Oracle* (www.oracle.com). Viewed 22 June 2007.

Ott T. and Swiaczny F. (2001) *Time-Integrative Geographic Information Systems: Management and Analysis of Spatio-temporal Data* Springer-Verlag: Berlin.

Pearson A. W. and Collier P. (1998) 'The Integration and Analysis of Historical and Environmental Data using a Geographical Information System: Landownership and Agricultural Productivity in Pembrokeshire c. 1850' *Agricultural History Review*, 46: pp. 162–176.

Pearson A. W. and Collier P. (2002) 'Agricultural History with GIS' in Knowles A. K. (ed.) *Past Time, Past Place: GIS for History* ESRI Press: Redlands, CA, pp. 105–116.

Peng Z.-R. and Tsou M.-H. (2003) *Internet GIS: Distributed Geographical Information Services for the Internet and Wireless Networks* John Wiley: Chichester.

Peuquet D. J. (1990) 'A Conceptual Framework and Comparison of Spatial Data Models' in Peuquet D. J. and Marble D. F. (eds.) *Introductory Readings in Geographic Information Systems* Taylor & Francis: London, pp. 250–285.

Peuquet D. J. (1994) 'It's About Time: a Conceptual Framework for the Representation of Temporal Dynamics in Geographic Information Systems' *Annals of the Association of American Geographers*, 84: pp. 441–461.

Peuquet D. J. (1999) 'Time in GIS and Geographical Databases' in Longley P. A., Goodchild M. F., Maguire D. J. and Rhind D. W. (eds.) *Geographical Information Systems. Volume I: Principles and Technical Issues* John Wiley & Sons: Chichester, pp. 91–103.

Pickles J. (1995a) 'Representations in an Electronic Age: Geography, GIS, and Democracy' in Pickles J. (ed.) *Ground Truth: the Social Implications of Geographic Information Systems* Guilford Press: New York, NJ, pp. 1–30.

Pickles J. (1995b, ed.) *Ground Truth: the Social Implications of Geographic Information Systems* Guilford Press: New York, NJ.

Pickles J. (1999) 'Arguments, Debates, and Dialogues: the GIS-social Theory Debate and the Concern for Alternatives' in Longley P. A., Goodchild M. F., Maguire D. J. and Rhind D. W. (eds.) *Geographical Information Systems: Principals, Techniques, Management and Applications*, 2nd edition Chichester: John Wiley, pp. 49–60.

Pickles J. (2004) *A History of Spaces: Cartographic Reason, Mapping and the Geocoded World* Routledge: London.

Pitternick A. (1993) 'The Historical Atlas of Canada: the Project behind the Product' *Cartographica*, 30: pp. 21–31.

Plewe B. (2002) 'The Nature of Uncertainty in Historical Geographical Information' *Transactions in GIS*, 6: pp. 431–456.

Polis Center (2001) *North American Religion Atlas* (www.religionatlas.org). Viewed 22 June 2007.

Raafat H., Yang Z. and Gauthier D. (1994) 'Relational Spatial Topologies for Historical Geographical Information' *International Journal of Geographical Information Systems*, 8: pp. 163–173.

Raper J. (2001) *Multidimensional Geographic Information Science* Taylor & Francis: London.

Ray B. C. (2002a) http://etext.lib.virginia.edu/salem/witchcraft. Viewed 19th December 2005.

Ray B. C. (2002b) 'Teaching the Salem Witchcraft Trials' in Knowles A. K. (ed.) *Past Time, Past Place: GIS for History* ESRI Press: Redlands, CA, pp. 19–33.

Reibel M. and Bufalino M. E. (2005) 'Street-weighted Interpolation Techniques for Demographic Count Estimation in Incompatible Zone Systems' *Environment and Planning A*, 37: pp. 127–39.

Reid J. S., Higgins C., Medyckyj-Scott D. and Robson A. (2004) 'Spatial Data Infrastructures and Digital Libraries: Paths to Convergence' *D-Lib Magazine*, 10 (5) (www.dlib.org). Viewed 22 June 2007.

Rey S. J. and Janikas M. V. (2006) 'STARS: Space-Time Analysis of Regional Systems' *Geographical Analysis*, 38: pp. 67–86.

Richards J. A. (1993) *Remote Sensing Digital Image Analysis: an Introduction* Springer-Verlag: Berlin.

Riley D. N. (1996) *Aerial Archaeology in Britain* Shire Publications: Princes Risborough.

Robinson A. H., Morrison J. L., Muehrcke P. C., Kimerling A. J. and Guptill S. C. (1995) *Elements of Cartography*, 6th edition John Wiley & Sons: Chichester.

Robinson J. M. and Zubrow E. (1997) 'Restoring Continuity: Exploration of Techniques for Reconstructing the Spatial Distribution underlying Polygonized Data' *International Journal of Geographical Information Science*, 11: pp. 633–648.

Robinson W. (1950) 'Ecological Correlations and the Behavior of Individuals' *American Sociological Review*, 15: pp. 351–357.

Roblin H. S. (1969) *Map Projections* Edward Arnold: London.

Robson B. T. (1969) *Urban Analysis* Cambridge University Press: Cambridge.

Rorty R. (1979) *Philosophy and the Mirror of Nature* Princeton University Press: Princeton, NJ.

Rumsey D. and Williams M. (2002) 'Historical Maps in GIS' in Knowles A. K. (ed.) *Past Time, Past Place: GIS for History* ESRI Press: Redlands, CA, pp. 1–18.

Sack R. D. (1972) 'Geography, Geometry and Explanation' *Annals of the Association of American Geographers*, 62: pp. 61–78.

Sack R. D. (1973) 'A Concept of Physical Space in Geography' *Geographical Analysis*, 5: pp. 16–34.

Sack R. D. (1974) 'Chronology and Spatial Analysis' *Annals of the Association of American Geographers*, 64: pp. 439–452.

Sadahiro, Y. (2000) 'Accuracy of Count Data Transferred through the Areal Weighting Interpolation Method' *International Journal of Geographical Information Sciences*, 14: pp. 25–50.

Saville J. (1957) *Rural Depopulation in England and Wales, 1851–1951* Routledge & Kegan Paul: London.

Schaefer M. (2003) 'Visually Interpreting History – the Battle of Hastings'. Paper presented at the 2003 Social Science History Association conference.

Schaefer M. (2004) 'Design and Implementation of a Proposed Standard for Digital Storage and Internet-based Retrieval of Data from the Tithe Survey of England and Wales' *Historical Methods*, 37: pp. 61–72.

Schofield R. (1965) 'The Geographical Distribution of Wealth in England, 1334–1649' *Economic History Review: Second Series*, 18: pp. 483–510.

Scollar I., Tabbagh A., Hesse A. and Herzog I. (1990) *Archaeological Prospecting and Remote Sensing* Cambridge University Press: Cambridge.

Shaw G. and Wheeler D. (1994) *Statistical Techniques in Geographical Analysis* David Fulton: London.

Sheehan-Dean A. C. (2002) 'Similarity and Difference in the Antebellum North and South' in Knowles A. K. (ed.) *Past time, Past Place: GIS for History* ESRI Press: Redlands, CA, pp. 35–50.

Shennan I. and Donoghue D. N. M. (1992) 'Remote Sensing in Archaeological Research' *Proceedings of the British Academy*, 77: pp. 223–232.

Shepherd I. D. H. (1995) 'Putting Time on the Map: Dynamic Displays in Data Visualization' in Fisher P. F. (ed.) *Innovations in GIS 2* Taylor & Francis: London, pp. 169–188.

Siebert L. (2000) 'Using GIS to Document, Visualize, and Interpret Tokyo's Spatial History' *Social Science History*, 24: 537–574.

Skinner G. W., Henderson M. and Jianhua Y. (2000) 'China's Fertility Transition through Regional Space' *Social Science History*, 24: pp. 613–652.

Smith N. (1992) 'History and Philosophy of Geography: Real Wars, Theory Wars' *Progress in Human Geography*, 16: pp. 257–271.

Smith D. M., Crane G. and Rydberg-Cox, J. (2000) 'The Perseus Project: a Digital Library for the Humanities' *Literary and Linguistic Computing*, 15: pp. 15–25.

Social Explorer (2006) *Social Explorer* (www.socialexplorer.org). Viewed 22 June 2007.

Southall H. R. (2006) 'A Vision of Britain through Time: Making Sense of 200 years of Census Reports' *Local Population Studies*, 76: 76–84.

Spence C. (2000a) *London in the 1690s: a Social Atlas* Centre for Metropolitan History, Institute of Historical Research: London.

Spence C. (2000b) 'Computers, Maps and Metropolitan London in the 1690s' in Woollard M. (ed.) *New Windows on London's Past: Information Technology and the Transformation of Metropolitan History* Association for History and Computing: Glasgow, pp. 25–46.

Steers J. A. (1950) *An Introduction to the Study of Map Projections*, 8th edition University of London Press: London.

Taylor P. J. (1990) 'Editorial Comment: GKS' *Political Geography Quarterly*, 9: pp. 211–212.

Taylor P. J. and Johnston R. J. (1995) 'GIS and Geography' in Pickles J. (ed.) *Ground Truth: the Social Implications of Geographic Information Systems* Guilford Press: New York NJ, pp. 51–67.

Thomas W. G. and Ayers E. L. (2003) 'An Overview: the Differences Slavery Made: a Close Analysis of Two American Communities' *American Historical Review*, 108: pp. 1298–1307, See also www.historycooperative.org/ahr/elec-projects.html. Viewed 22 June 2007.

TimeMap (2005) *TimeMap: Time-based Interactive Mapping* (www.timemap.net). Viewed 22 June 2007.

Tobler W. (1970) 'A Computer Movie Simulating Urban Growth in the Detroit Region' *Economic Geography*, 46: pp. 234–240.

Tomlin C. D. (1991) 'Cartographic Modelling' in Maguire D. J., Goodchild M. F. and Rhind D. W. *Geographical Information Systems: Principles and applications. Volume I: Principles* Longman Scientific and Technical: London, pp. 361–374.

Tomlinson R. F. (1990) 'Geographical Information Systems – a New Frontier' in Peuquet D. J. and Marble D. F. (eds.) *Introductory Readings in Geographic Information Systems* Taylor & Francis: London, pp. 18–29.

Townsend S., Chappell C. and Struijvé O. (1999) *Digitising History: a Guide to Creating Digital Resources from Historical Documents* Oxbow Books: Oxford.

Tufte E. R. (1990) *Envisioning Information* Graphics Press: Cheshire, CT.

Tukey J. W. (1977) *Exploratory Data Analysis* Addison-Wesley: Reading, MA.

Unwin D. (1995) 'Geographic Information Systems and the Problem of "Error and Uncertainty"' *Progress in Human Geography*, 19: pp. 549–558.

Unwin D. (1996) 'Integration through Overlay Analysis' in Fischer M., Scholten H. J. and Unwin D. (eds.) *Spatial Analytical Perspectives on GIS* Taylor & Francis: London pp. 129–138.

US Bureau of the Census (1990) 'Technical Description of the DIME System' in Peuquet D. J. and Marble D. F. (eds.) *Introductory Readings in Geographical Information Systems* Taylor & Francis: London, pp. 100–111.

VanHaute E. (2005) 'The Belgium Historical GIS' *Historical Geography*, 33: pp. 136–139.

Verbyla D. L. (2002) *Practical GIS Analysis* Taylor & Francis: London.

Vision of Britain (2005) *A Vision of Britain through Time* (www.visionofbritain.org.uk). Viewed 22 June 2007.

Visvalingham M. (1994) 'Visualisation in GIS, Cartography and ViSC' in Hearnshaw H. M. and Unwin D. J. (eds.) *Visualisation in GIS* John Wiley & Sons: Chichester, pp. 18–25.

Vrana R. (1989) 'Historical Data as an Explicit Component of Land Information Systems' *International Journal of Geographical Information Systems*, 3: pp. 33–49.

Vrana R. (1990) 'Historical Data as an Explicit Component of Land Information Systems' in Peuquet D. J. and Marble D. F. (eds.) *Introductory Readings in Geographical Information Systems* Taylor & Francis: London, pp. 286–302.

Wachowicz M. (1999) *Object-orientated Design for Temporal GIS* Taylor & Francis: London.

Walford N. (2005) 'Connecting Historical and Contemporary Small-area Geography in Britain: the Creation of Digital Boundary Data for 1971 and 1981 Census Units' *International Journal of Geographical Information Science*, 19: pp. 749–767.

Weibel R. and Heller, M. (1991) 'Digital Terrain Modelling' in Maguire D. J., Goodchild, M. F. and Rhind D. W. (eds.) *Geographical Information Systems: Principles and Applications. Volume I: Principles* Longman Scientific and Technical: Harlow, pp. 269–297.

Wilson A. (2001) 'Sydney TimeMap: Integrating Historical Resources Using GIS' *History and Computing*, 13: pp. 45–69.

Wood M. and Brodlie K. (1994) 'ViSC and GIS: Some Fundamental Considerations' in Hearnshaw H. M. and Unwin D. J. (eds.) *Visualisation in GIS* John Wiley & Sons: Chichester, pp. 3–8.

Woods R. (1979) *Population Analysis in Geography* Longman: London.

Woods R. and Shelton N. (1997) *An Alas of Victorian Mortality* Liverpool University Press: Liverpool.

Woollard M. (2000, ed.) *New Windows on London's Past: Information Technology and the Transformation of Metropolitan History* Association for History and Computing: Glasgow.

Worboys M. (1998) 'A Generic Model for Spatio-bitemporal Geographic Information' in Egenhofer M. and Golledge R. (eds.) *Spatial and Temporal Reasoning in Geographic Information Systems* Oxford University Press: Oxford, pp. 25–39.

Worboys M. F. (1999) 'Relational Databases and Beyond' in Longley P. A., Goodchild M. F., Maguire D. J. and Rhind D. W. (eds.) *Geographical Information Systems: Principals, Techniques, Management and Applications*, 2nd edition John Wiley: Chichester, pp. 373–384.

Worcester D. (1982) *Dust Bowl: the Southern Plains in the 1930s* Oxford University Press: Oxford.

Wrigley N., Holt T., Steel D. and Tranmer M. (1996) 'Analysing, modeling, and resolving the Ecological Fallacy' in Longley P. and Batty M. (eds.) *Spatial Analysis: Modeling in a GIS Environment* GeoInformation International: Cambridge, pp. 25–40.

Zadeh, L. (1965) 'Fuzzy Sets' *Information and Control*, 8: pp. 338–353.

INDEX

Printed in the United States
By Bookmasters